EXTREME ENVIRONMENTS
Mechanisms of
Microbial Adaptation

Academic Press Rapid Manuscript Reproduction

EXTREME ENVIRONMENTS
Mechanisms of
Microbial Adaptation

EDITED BY
Milton R. Heinrich

Biological Adaptation Branch
Ames Research Center
National Aeronautics and Space Administration
Moffett Field, California

Academic Press, Inc.
New York   San Francisco   London   1976
A Subsidiary of Harcourt Brace Jovanovich, Publishers

ACADEMIC PRESS, INC.
111 Fifth Avenue, New York, New York 10003

United Kingdom Edition published by
ACADEMIC PRESS, INC. (LONDON) LTD.
24/28 Oval Road, London NW1

Library of Congress Cataloging in Publication Data
Main entry under title:

Extreme environments.

    Papers of a symposium held at Ames Research
Center, Moffett Field, Calif., June 26-28, 1974;
sponsored by the Biological Adaptation Branch of the
center
    Bibliography: p.
    Includes index.
    1.   Micro-organisms—Physiology—Congresses.
2.   Adaptation (Physiology)—Congresses.   3.   Microbial ecology—Congresses.   I.   Heinrich, Milton R.
II.   United States.   Ames Research Center, Moffett
Field, Calif.   Biological Adaptation Branch.

QR97.A1E93   576'.15   75-38719
ISBN 0-12-337850-8
PRINTED IN THE UNITED STATES OF AMERICA

Dr. Wolf V. Vishniac
with an early model of the "Wolf Trap"

Contents

PROTEINS

MEMBRANES

List of Contributors

M. Alexander, Laboratory of Soil Microbiology, Department of Agronomy, Cornell University, Ithaca, New York

Stanley Bayley, Department of Biology, McMaster University, Hamilton, Ontario, Canada

Roy E. Cameron, Darwin Research Institute, Dana Point, California[†]

Alfred F. Esser, Department of Chemistry California State University, Fullerton, California

Roger D. Haight, Department of Biological Sciences, San Jose State University, San Jose, California

Lawrence I. Hochstein, Biological Adaptation Branch, NASA, Ames Research Center, Moffett Field, California

Richard C. Honour, Darwin Research Institute, Dana Point, California

Harold P. Klein, Director of Life Sciences, NASA, Ames Research Center, Moffett Field, California

Janos K. Lanyi, Biological Adaptation Branch, NASA, Ames Research Center, Moffett Field, California

Lars G. Ljungdahl, Department of Biochemistry, University of Georgia, Athens, Georgia

Robbe C. Lyon, Program in Biochemistry and Biophysics, Washington State University, Pullman, Washington

R.D. MacElroy, Biological Adaptation Branch, NASA, Ames Reaearch Center, Center, Moffett Field, California

James A. Magnuson, Program in Biochemistry and Biophysics, Washington State University, Pullman, Washington

[†]Present address: Argonne National Laboratory, Argonne, Illinois

Nancy S. Magnuson, Program in Biochemistry and Biophysics, Washington State University, Pullman, Washington

Ronald N. McElhaney, Department of Biochemistry, University of Alberta, Edmonton, Alberta, Canada

C. R. Middaugh, Biological Adaptation Branch, NASA, Ames Research Center, Moffett Field, California

Frank A. Morelli, Darwin Research Institute, Dana Point, California

Brian R. Reid, Department of Biochemistry, University of California, Riverside, California

David Sherod, Department of Biochemistry, University of Georgia, Athens, Georgia

Rivers Singleton, Jr., Biological Adaptation Branch, NASA, Ames Research Center, Moffett Field, California[†]

Kenneth A. Souza, Biological Adaptation Branch, NASA, Ames Research Center, Moffett Field, California

J. Stenesh, Department of Chemistry, Western Michigan University, Kalamazoo, Michigan

Gabor Szabo, Department of Physiology, University of California, Los Angeles, California

James R. Trudell, Department of Anesthesiology, Stanford University School of Medicine, Stanford, California

Ian L. Uydess, Department of Biology, University of Rochester, Rochester, New York

Wolf V. Vishniac, Department of Biology, University of Rochester, Rochester, New York[*]

Neil E. Welker, Department of Biochemistry and Molecular Biology, Northwestern University, Evanston, Illinois

[†] Present address: Department of Biological Sciences, University of Delaware, Newark, Delaware
[*] Deceased December 10, 1973.

Preface

The ability of microorganisms to adapt to hostile environments of every description has intrigued generations of microbiologists. The versatility of these organisms has enabled them to populate every type of stable environment on Earth, including hot springs, the polar regions, deserts, salt ponds, acid or alkaline waters, and many others. In recent years it has been realized that these organisms can be important tools for probing many areas of microbial physiology. They offer examples of changes in molecular structure, metabolic pathways, or structural elements which can be related directly to an external parameter. They also extend the range of conditions over which functioning cellular constitutents can be studied. In addition, with the increased emphasis on the environment, it is now appreciated that these organisms are important factors in the Earth's ecology, and their potential for solving some of the environmental problems of the future should not be overlooked.

With the beginning of planetary exploration, there was increasing interest in the possibility of life on the other planets. As information about their environments accumulated, it became evident that many of the conditions were no more severe than some of those found on Earth, and to which terrestrial organisms had successfully adapted. It appeared that the study of the processes of adaptation would provide information of use in considering the possibilities for the occurrence of native life on a planet, or the probability of survival of terrestrial microorganisms inadvertently introduced on another planet.

The laboratory within the National Aeronautics and Space Administration that is responsible for research in the adaptation of microorganisms to extreme environments is the Biological Adaptation Branch of the Ames Research Center, Moffett Field, California. This laboratory has recognized the need for better communication among workers in the area of microbial adaptation. Although the environments being studied are diverse, there are common principles, approaches, and techniques. Insights gained in one area can be useful in other areas. In addition, there is the obvious need to correlate current information in an area, and this in turn will suggest future directions for the work. The need is especially urgent in this field, where the pertinent literature is scattered among many journals of both microbiology and biochemistry. For these reasons, the branch has organized a series of biennial meetings under the title "Extreme Environments: Mechanisms of Microbial Adaptation." The emphasis has not been on descriptive microbiology, but on work leading to an understanding of the molecular mechanisms that have allowed the organism to survive. The first meeting, in 1970, provided a review of the state of knowledge of adaptation to most of the specific environments which had received appreciable study. The second meeting (1972) examined in more detail three topics closely related to adaptation: proteins, membranes, and autotrophy. The proceedings of the first two meetings were not published, but their programs are given in the Appendix.

The papers presented at the 1974 meeting are published in this volume. Following a Theme Lecture on general aspects of microbial adaptation, by Martin Alexander, papers were presented on the effects of temperature and high salt concentration on the various steps in information transfer and on microbial enzymes. Also examined were the effects of temperature, salt, and pressure on membrane structure and function.

While planning for the meeting was in progress, this community was saddened by the news of the tragic death of Dr. Wolf Vishniac in the Antarctic. He was a personal friend to many workers in this field, and respected by all. It was his desire personally to collect the samples from that extreme environment which led to his death. It seemed appropriate that this meeting should be dedicated to Wolf Vishniac. A dedication was presented by H. P. Klein, a friend of many years; also included were a paper on work in progress at the time in the Vishniac laboratory, by his associate Ian Uydess, and a review of Antarctic microbiology by Roy Cameron.

The Editor expresses his deepest gratitude not only to the contributors and publisher, but to the members of the Biological Adaptation Branch for their part in the planning and execution of the meeting, and to Dr. Richard S. Young, Chief of the Planetary Biology Office of NASA, for financial support.

DR. WOLF V. VISHNIAC

Harold P. Klein

This conference is dedicated to the memory of Wolf Vishniac. Those of you who knew Wolf will understand why it is almost inappropriate for me to refer to him as *Doctor* Wolf Vishniac. For the others, let me say simply that Wolf was one of the kindest, gentlest, most unaffected human beings I have known.

Our paths — Wolf's and mine — go back to a common point almost 35 years ago. And these paths have intertwined many, many times, since then.

I first met Wolf when we were both undergraduate students at Brooklyn College, in the early 1940s. He had been born in Berlin, of Latvian parentage, in 1922, and came to America (by way of Sweden) in 1940. At Brooklyn College, he showed an unusual aptitude and enthusiasm for biology. This, coupled with his easy ability to communicate with people, resulted in his being elected to the presidency of the undergraduate biomedical society — an organization of major proportions on the campus.

I think it was back there — at Brooklyn College — that Wolf first became fascinated by the subject of microbial physiology, in courses with Carrol Grant and James Weiss. He went on from Brooklyn, by way of a Masters degree at Washington University, to Stanford University, where he became the doctoral student of "Kase" Van Niel. From 1946 to 1949, he studied with Van Niel at the Hopkins Marine Station in Pacific Grove. Again during those years, our paths touched, for I was studying at the University of California in Berkeley with Doudoroff and Stanier at the same time, and we met several times—in Berkeley, at Asilomar or in Pacific Grove — exercising the friendly rivalry that existed between the two groups. Wolf, on those occasions, talked only of research; of the beauty of truth. For Wolf, the future was to seek the truth.

Then, upon receiving his Ph.D. degree in 1949, he moved on to New York, to the laboratory of Severo Ochoa. Once again, fate played her little game, as I went to Boston at about the same time, to Ochoa's major competitor — Fritz Lipmann. I did not see much of Wolf during those years. I suspect that Ochoa may have driven his people harder than Lipmann did.

Whatever happened, Wolf made his place in the history of biology there; he was the first to show, with Ochoa, that there was a photochemical reduction of pyridine nucleotide concomitant with the fixation of CO_2. In further elaboration of these studies, Wolf showed that phosphorylation was coupled to the light-dependent reduction of TPN. These observations established Wolf as a major contributor to the fundamental store of information on plant photosynthesis.

Following his fellowship with Ochoa, Wolf moved to the Microbiology Department of Yale University, where, for the next 10 years, he concentrated his research on the metabolism of the autotrophic bacteria. He made notable contributions to our knowledge of *Thiobacillus* and *Hydrogenomonas*, and he made occasional scientific forays into the physiology of algae, molds, and higher plants.

It may have been his concern with what were then regarded as "exotic" organisms — the autotrophs — which led him into the field of exobiology. Perhaps, wondering about the diversity and adaptability of terrestrial life, Wolf found it easier than the rest of us to think seriously in terms of extraterrestrial life.

By 1959, Wolf had become an active charter member of the elite fraternity of exobiologists, along with Lederberg, Sagan, and others. He became an ardent and eloquent spokesman for his fledgling Science, and over the ensuing years assumed a major role as a proponent of this subject. The list of committees, panels, and boards that he chaired, or on which he served, is staggering. He was the leading exponent of exobiology on NASA's prestigious Lunar and Planetary Missions Boards; he became chairman of COSPAR's biology working group; he was a key member of the Space Sciences Board of the National Academy of Sciences; he wrote and lectured tirelessly on the subject of exobiology.

From 1966 through 1969, I had the pleasure of Wolf's company again on the Interagency Committee on Back-Contamination. Along with representatives from the U.S. Public Health Service, Department of Agriculture, Department of Interior, and others, this committee was charged with setting the requirements for lunar quarantine on the Apollo missions. It was Wolf who kept insisting on the highest standards in developing and maintaining these procedures. He was so deeply concerned with these matters that he helped train astronauts by organizing and conducting courses in Houston on the biological implications of handling the lunar material.

In 1961, Wolf left Yale, and went to the University of Rochester. There, as Professor of Biology, he continued his investigations on the metabolism of sulfur in *Thiobacillus*. But his main technical interest inexorably turned to the planet Mars. He was one of the first people in this country to conceive of and develop an automated system to measure microbial metabolism on Mars. This concept — dubbed the "Wolf Trap" — was simply a system to measure microbial turbidity after inoculating a suitable medium with soil. The concept was developed to a high level of perfection, and Wolf's experiment was selected to be one of the four biological experiments to be flown on the Viking Mars mission in 1975. In the Viking version, Martian soil was to be incubated under Martian atmosphere in the presence of simply distilled water, and the incubation chamber was to be monitored by a sensitive optical system.

Again in this endeavor, Wolf and I came to work closely together. Both of us were selected to be on the seven-man Viking Biology Team; Wolf became the deputy team leader. In addition to developing his own experiment, he was our main interface with other science teams on the complex Viking mission.

But in 1971, in a fever of financial retrenchment, NASA removed Wolf's experiment from the Viking payload. To say that he was disappointed, after 12 years of commitment and hard work in support of planetary exploration, would be a gross understatement. Had he "slammed the door," and stormed out of our lives, we would not have been surprised. But this was not Wolf's style! After accepting this decision, Wolf came back and asked, "What can I do? How can I help the mission?"

He decided that he could contribute most richly to the mission by studying the dry valleys of Antarctica. Here the environment is as close as we could get to that on Mars. Cameron, Horowitz, and colleagues had reported that these valleys were devoid of life, but the microbiologist in Wolf found this hard to accept.

So in 1971–1972, he went off on the first of his two trips, and returned with data contrary to the earlier reports. Wolf had evidence for the prevalence of a sustained microbial population in this very harsh environment. He was sorely disappointed when his second projected trip to Antarctica, to extend his earlier work, was delayed at one point. But he telephoned me in the fall of 1973 for the last time and joyously informed me that all the red tape had been eliminated and that he was imminently going to make another study in Antarctica. This was in September.

He died there on December 10, 1973, while out collecting soil samples. He apparently fell down a slope and over a cliff on the slopes of Upper Wright Valley. We lost a very valuable scientific colleague and the world lost a beautiful human being.

EXTREME ENVIRONMENTS
Mechanisms of
Microbial Adaptation

THEME LECTURE

NATURAL SELECTION AND THE ECOLOGY OF MICROBIAL ADAPTATION IN A BIOSPHERE

M. Alexander

An enormous literature exists on natural selection operating among populations of higher animals and plants, and similarly, there is a vast body of information on microbial adaptation in individual populations of bacteria and fungi, *in vitro* at any rate. However, with the notable exception of a fascinating essay by van Niel (1955), natural selection in microbial communities of soils and waters is a largely ignored subject. Furthermore, speculation on the potential impact of agents of selection in determining the composition of the biosphere of other planets is almost wholly lacking.

That natural selection functions to govern the composition of existing communities is, so many years after Darwin, now really no more than a truism, but to anticipate how selective forces might lead to the proliferation of novel genotypes and phenotypes on other planets of this or other solar systems is a topic of some relevancy as man seeks for new living systems on Mars and, at some future date, even more distant celestial bodies.

From the viewpoint of the putative Martian, the appearance of life on earth must seem to have been associated with a selection of organisms adapted to truly extreme conditions. The Earth creature must often be able to withstand enormous quantities of water; be capable of living in a region with high atmospheric pressures; and, most frightening of all, be able to develop means of overcoming or withstanding the toxic effects of that highly poisonous substance, molecular oxygen. From the areocentric viewpoint, the selective pressures must have been enormous to allow for the evolution of species able to cope with these harsh circumstances. After all, the Martian Pangloss--Voltaire's incurable optimist--must believe that all is for the best in his best of all possible worlds, one containing little oxygen, considerable carbon dioxide, little water, and an atmosphere with no enormous overburdens of pressure.

We geocentric biologists must examine the strategy of natural selection, evolution, and adaptation on neighboring planets in much the same way that extraterrestrial biologists would view the earth. Natural selection is a powerful force to model communities to fit the prevailing circumstances, provided, of course, that the spark of life has been kindled and that self-replicating systems have a chance of surviving and multiplying. These molding forces dictate that the indigenous organisms be capable of coping with their surroundings and they will favor species bearing fitness traits that allow their possessors to multiply, endure the physical and chemical stresses of their habitat, and make use of appropriate energy sources, nutrients, and electron acceptors in their surroundings. Individuals and populations having these fitness traits would assume prominence, whereas those not so favorably endowed would become minor members of the extraterrestrial community, or would disappear completely. The species would then have those allegedly novel characteristics that permit any organism to multiply in its native surroundings. Much as the Earthlings find no difficulty in growing in an environment containing a superabundance of water, levels of oxygen that might appear to be lethal, and an atmosphere that might eliminate alien species, so the indigenous forms of other biospheres would have all the necessary fitness traits to permit colonization of their native environments.

The limitations of space prevent my dealing in any detail with natural selection in microbial communities and microbial adaptation. However, problems of interspecific and intraspecific selection in microbial communities and examples based upon experience with free-living and parasitic micro-organisms have been reviewed (Alexander, 1971). I shall not endeavor to enter into a discussion of the mechanisms of adaptation, the means by which individuals or populations are modified to allow them to be more in harmony with new environments. Instead, I would like to present an ecological overview of stresses in extreme terrestrial and aquatic environments in this biosphere, and to comment briefly on the ecology of adaptation to environmental stress. The approach is that of the microbial ecologist, and not of the specialist seeking to understand the basic mechanisms of adaptation in individuals or populations or the underlying forces in natural selection among representatives of a community.

Although clearly defined studies are not abundant, and illustrations among microorganisms are few, it seems reasonable to conclude, from the available information, that species diversity declines as environmental adversity increases. Thus, the greater the severity of some stress factor imposed upon a natural habitat, the fewer the number of species that are present. Possibly, the populations that assumed dominance were always present in the original nonharsh habitat, and only became evident as the detrimental factor became more prominent, or alternatively, the organism was indigenous to an entirely different habitat and was transported to the new site by one or another of the many means of dissemination. Evidence for such low species diversity in extreme conditions is evident in polluted waters, the hot springs that have been so actively studied in recent years, algal communities of alpine habitats, regions of very low or extremely high pH, or areas receiving high concentrations of antimicrobial chemicals used in agricultural practice or present in industrial wastes discharged into waterways. An excellent illustration of the effect of extremes of acidity and temperature on the number of species and varieties of blue-green algae in Yellowstone National Park is shown in Figure 1.

STRESSES IN EXTREME ENVIRONMENTS ON EARTH

Let us turn to some of the environments that are so far from the normal on earth that they are deemed extreme, the inhabitants having been either selected from among arrivals in that locality able to cope with the prevailing stresses, derived from an indigenous population containing individuals that adapted to the stress factor, or always able to tolerate it.

A stress factor of common ecological interest is salinity, and several investigators have characterized populations in waters and soils containing high concentrations of salts. The few microbial species reported in these localities implies a low biological diversity. Thus, only a few algae have been found in Great Salt Lake, including species of Coccochloris

5

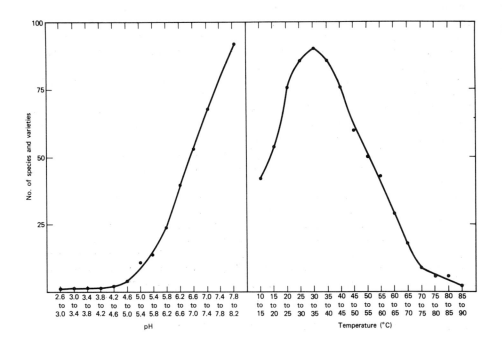

Fig. 1. *Distribution of blue-green algae as related to*
environmental pH and temperature. *From Copeland (1936)*

among the blue-greens, and of Chlamydomonas among the greens.
A number of protozoa have also been encountered in the lake
(Flowers and Evans, 1966). From a report on the inhabitants
of the Dead Sea, it is evident that red halophilic bacteria
and the green flagellated alga, Dunaliella viridis, are
dominant members of the community (Kaplan and Friedmann, 1970),
such organisms being typical of extremes of salinity. Solar
salt works, in which the salt level becomes progressively
higher in succeeding ponds, represent another extreme milieu
for microscopic life. In such habitats, Dunaliella,

Coccochloris, and a few other uniquely adapted algae are
frequently noted (Davis, 1968).

Some African lakes are even richer in salt than the Dead
Sea. In one such body of water, in which the salt content
was about 400 gm/liter, a surprisingly large array of cocci,
vibrios, and asporogenous and spore-forming bacteria were
noted (Brisou *et al.*, 1974). Clearly, the highly saline
waters restrict the abundance of physiological types, as is
evident from the paucity of obligate anaerobes other than
sulfate-reducing bacteria. Bacteria of various sorts have also
been obtained from saline soils and brines, and many of these
fail to grow in salt-poor solutions. The latter observation
is consistent with the frequent finding that a microorganism
adapted to a particular stress circumstance often fails to
grow at a lower intensity of the same factor. In effect, the
organism has purchased the virtue of growing in one area at
the expense of its ability to grow in a second. The same
inability to grow in conventional media must be borne in mind
in any system for detecting life on other planets: The
tolerance range or ecological amplitude of the alien might
not encompass the range of conditions included in the media
designed for the life-detection system. The biochemical basis
for extreme halophilism, of course, has been the focus of many
investigations.

A common stress in soils is drought. Land, particularly
under a partial plant canopy, is subject to regular cycles of
wetting and drying, and a population must be able to endure
these fluctuations if it is to maintain itself in the
subterranean habitat. That so many bacteria, actinomycetes,
and fungi do so attests to the fact that regular cycles of
wetting and drying do not constitute an extreme condition.
But undoubtedly many potential colonizers, for example, numerous
algae, fail to become established because of their inability
to endure even brief periods of desiccation.

Prolonged drought, on the other hand, has a devastating
impact. This is illustrated by observations made during a
severe drought that occurred in Kenya. Before the drought,
the actinomycetes accounted for 30% of the viable propagules
in the soil, but they reached a value of 90% of the total
culturable organisms as the drought progressed (Meiklejohn,
1957), their increasing abundance apparently resulting from
the resistance of their conidia to inactivation by drying.
A similar resistance of the actinomycetes, largely Streptomyces,

TABLE I

Survival of microorganisms in Desiccated Collamer Silt Loam[†]

	Counts per gram $\times 10^6$				
Days	Total	Actino-mycetes	Spore Formers	Arthro-bacter	Nonsporing Rods	Cocci
0	43	18	7.0	6.0	9.0	3.0
7	4.6	2.1	1.2	0.4	0.5	0.4
60	2.7	1.5	1.0	0.1	0.1	0.0
330	0.7	0.2	0.5	0.0	0.0	0.0

[†]
From Chen and Alexander (1973).

to desiccation has been observed in the laboratory (Table 1), the data also demonstrating the expected longevity of sporeformers of the genus <u>Bacillus</u> and the surprising durability of members of the non-sporeforming genus <u>Arthrobacter</u>. These data have practical implications inasmuch as they suggest that the physiological versatility characteristic of the soil microflora, and so important to plant growth, may be diminished following long periods without rain.

The water relationships of fungi, yeasts, and heterotrophic bacteria are of special importance in the food industry, owing to the use of sugar and salt as preservatives and to the appearance of spoilage organisms in syrups, honey, dried fruits, and salted meats. Limiting concentrations for growth of some of these organisms may be 20% NaCl or 80% sucrose, and some of the fungi will even germinate at water activities (a_w) of 0.605 (Onishi, 1963; Pitt and Christian, 1968). The ranges of microbial tolerance to moisture levels or salt and sugar

concentrations have been extensively explored because of
interest in the food industry, and some attention likewise
has been given to the origin and mechanisms of adaptation of
these unique organisms. The truism that species diversity
declines as the intensity of the harsh environmental
condition increases is, in more prosaic terms, well known to
the housewife, who often finds a nearly axenic culture in a
jar of jelly or in a bottle of maple syrup. The principles
derived from even these prosaic observations--the adaptability
of microorganisms, their capacity to endure adversity, and
the homogeneity of the resulting community--have considerable
relevancy to ecology and physiology, not to mention planetary
exploration.

Sites of high temperature are ever present on the Earth's
surface, and the species inhabiting these odd habitats and the
biochemical basis of the adaptation have long attracted
microbiologists. Thermophiles are encountered frequently in
decomposing animal manure, composts made from crop remains,
and other accumulations of carbonaceous materials, and the
genera active in such thermal habitats are well documented.
Of particular note in such heating heaps of organic matter
are the actinomycetes, a group rarely observed as dominants
in degradative processes, but astonishingly little work has
been done on these thermophiles. One must single out T.D.
Brock and his associates for providing fascinating insights
into the more extreme of the thermal habitats, in particular,
the hot springs. As a result of these studies, the ecology
of blue-green algae of the hot springs of Yellowstone National
Park and the unique sulfur bacteria that are able to metabolize
at temperatures of 93° C is now well known (Brock, 1970; Brock
et al., 1971). On the basis of a number of observations,
Tansey and Brock (1972) proposed that the upper temperature
limit for eukaryotic organisms is near 60° C, a limit they
suggest may be related to the failure for evolution to give
rise to eukaryotes able to form functional organelle membranes
that are stable at higher temperatures.

By contrast with the lure of heat, the extremes of cold
have not attracted many microbiologists in recent years to
initiate physiological inquiry, although minimum temperatures
for replication of individual populations can be established
with ease. The lower limit for active life on earth is
probably in the vicinity of -18° C (Vallentyne, 1963).
Nevertheless, a distinct low-temperature microbiota exists,
and lists are available of the genera and often species of
algae, bacteria, fungi, protozoa, and lichens indigenous to

9

the Arctic and Antarctica. Here, too, it is common to find one or a few species to be dominant, and these to be abundant in the habitat. Wolf Vishniac was endeavoring to establish Antarctic microbiology on a more firm basis at the time of his tragic death.

Mountain climbing and alpine excursions are apparently more exciting than polar exploration to phycologists, and hence, the algae of mountain snows are more thoroughly documented. Alpine, glacial, and even polar snows bear a specialized group of algae that are frequently evident to the naked eye because of the striking colors they impart to the snow. Green, red, pink, orange, and yellow patches develop on the snow as a consequence of colonization by one or another of these algae, with species of Chlamydomonas, Ankistrodesmus, Mesotaenium, or Euglena often being responsible for the colors. The abundance of light in many permanent snow fields, and of water coming from the melting snows, are obvious features favorable to the algae, but the characteristic physiological attributes of the colonizers and the reasons for the dominance of individual species have yet to be defined. Stein (1968), however, reported that a Chlamydomonas associated with the snow was quite clearly a psychrophile, growing from 0-10°C.

Soils drained from offshore regions of the sea and certain volcanic lakes have a low pH and evidently support novel communities, but problems owing to pollution from mining operations have been the chief impetus to characterizing the biotas of highly acidic habitats. In the presence of oxygen, sulfide in iron and other ores is oxidized, and the resulting H_2SO_4 causes a precipitous fall of pH in the mine waters. Values of below 4.0 or even lower than 3.0 are commonly encountered. The impact on the aquatic biota is devastating, and entirely new populations come to the forefront. The survivors of the acid flows are few in number, but the biota is often quite exceptional. Protozoa and algae previously uncommon assume dominance in the waters (Lackey, 1939), and yeasts and fungi sometimes predominate among the aerobic heterotrophs (Tuttle *et al.*, 1968). Alkaline waters have received little attention, but a study of two lakes in Kenya with pH values above 10.0 showed the presence of a community typified by a dozen or fewer algal species (Jenkin, 1936).

An enormous portion of the biologically inhabited part of this planet is subjected to greater than one atmosphere of pressure, and the hydrostatic pressure in the oceans ranges from nearly one in surface waters to greater than 1000 atmospheres in some of the deeper zones. Many of the isolates from deep waters are incapable of replicating at atmospheric pressures, thus having a unique adaptation to their sea-bottom milieu, one that prevents culturing in conventional laboratory conditions. By contrast, some of the sea-bottom isolates can grow at one atmosphere (Morita, 1967), thereby posing an interesting question concerning the potential abilities of species from a biosphere of a certain atmospheric pressure to multiply on a second planet having different pressures at its surface. One might expect low pressures to have little influence on microbial activity, assuming that water is available in liquid form.

By virtue of its use of an astonishing array of toxic chemicals in industry and agriculture, human society is providing a whole series of new and extreme environments for biological investigation. Heavy metals from industry, organic chemicals that are toxic at the levels discharged into waters from factories, and fungicides used in crop production frequently cause catastrophic upsets in microbial communities, but the ecologist finds happy hunting grounds among the few survivors. Biological novelties have also been uncovered in the laboratory in tests involving assessment of the potential capacity of heterotrophs to proliferate in solutions containing high levels of toxicants; for example, a strain of <u>Bacillus pumilus</u> was found to grow in a saturated KCN solution (Skowronski and Strobel, 1969), and a gram-negative bacterium was capable of multiplying in an aqueous solution containing 45% (w/w) LiCl (Siegel and Roberts, 1966).

It is tempting to engage in an ecological-stress numbers game. Salinity and high pH often coexist, so the inhabitants of such sites must possess fitness traits allowing them to overcome both stresses. Acid-tolerant thermophiles have recently been isolated from hot acid soils and hot springs that have a low pH, and a bacterium like <u>Sulfolobus</u> will convert elemental sulfur to sulfuric acid at temperatures up to 85° C (Mosser et al., 1973). Similarly, a fungus like <u>Acontium</u> <u>velatum</u> is able to tolerate 4% $CuSO_4$ at pH values below 1.0 (Starkey and Waksman, 1943). Possibly no spectrum of harsh circumstances is as diverse as that to which are exposed the algae present on seashores between high and low tides. These organisms are desiccated at low tide, exposed to the constant mechanical action of waves, subject to rising salt levels as the sea water evaporates and to falling concentrations during a rain, exposed to high light intensities with essentially no protection, and frozen and thawed according to season in temperate latitudes. That they survive, no less grow, bears witness to their armament of fitness traits and to the forces of natural selection that have eliminated less versatile species.

ECOLOGY OF ADAPTATION TO ENVIRONMENTAL STRESS

Adaptation to prevailing conditions is evident not only in environments so harsh that few species survive, but also under more moderate conditions. For example, although sodium and chloride ions are not common nutrients for bacteria, they are required by, or markedly stimulate many marine bacteria. This suggests that an ecological advantage is acquired by populations having such nutrient requirements, provided of course that their surroundings also contain the ions. Similarly, algae in coastal waters, which receive iron-rich substances, have a high iron requirement, whereas those in the iron-poor open sea fare well at very low levels of the same element (Figure 2). In like fashion, acid soils have more acid-tolerant actinomycetes than do soils of higher pH values (Figure 3). Resistance among algae to cycles of freezing and thawing likewise reflects a natural selection, with those isolates from Antarctica that have been examined being consistently capable of surviving the temperature cycles, whereas only some of those from the temperate zone possess an analogous resistance (Holm-Hansen, 1967). A striking but similar type of adaptation, but to warm temperatures, is evident among certain protozoa. Thus, an amoeba that is to bring about primary amoebic meningoencephalitis must be able to survive the high fever temperatures developing in the host's body, and indeed, the pathogenic Naegleria fowleri can multiply at temperatures up to 44°C to 46°C, whereas Naegleria gruberi, a related but nonpathogenic protozoan, fails to grow above 37°C (Griffin, 1972).

The range of tolerance to many abiotic stresses can be increased in a given direction in vitro by repeated subculture of the test population in media with increasing intensities of the harsh factor. Rarely is it clear in the many acclimation studies that have been published whether the modification resulted from a genetic or a nongenetic change,

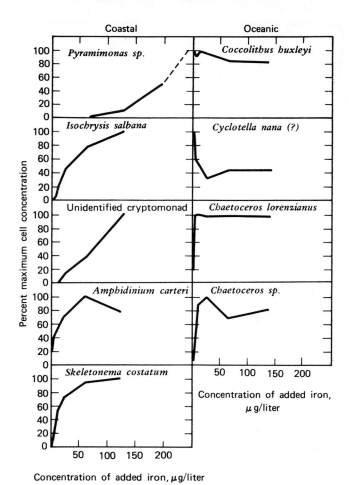

Fig. 2. Iron requirements of oceanic and coastal algae. Ryther and Kramer (1961).

and what was the precise cause of the change in genotype or phenotype. In typical illustrations, a mesophilic strain of <u>Bacillus</u> <u>subtilis</u>, with a maximum temperature for growth of 55° C, was acclimated to grow at 72° C by incubating serial transfers of the bacterial culture at progressively higher temperatures (Dowben and Weidenmuller, 1968). In the last century Dallinger (1887) was able to increase the temperature tolerance of several flagellated protozoa. One must assume that similar acclimations must have occurred among pathogens of warm-blooded animals, many of which have

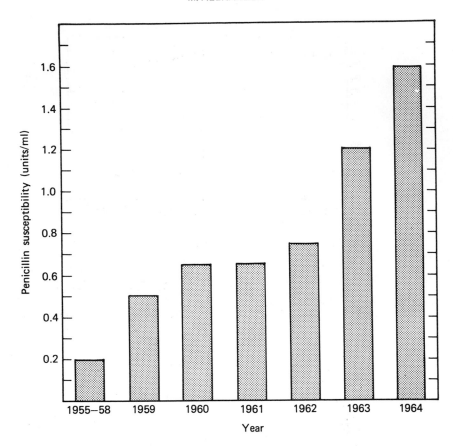

*Fig. 3.   Relative abundance of actinomycetes from two soils
able to grow at various pH levels in laboratory
media.   Corke and Chase (1964).*

optima in the vicinity of their host's body temperature.
Natural selection also takes place among populations for new
temperature regimes, so that organisms more suited to the
prevailing conditions assume dominance.   This is evident in
a study in which a mixed inoculum was incubated at 10, 45,
and 60°C to allow for selection; then the temperature range
for glucose metabolism by the inoculum was assessed (Figure 4).

Acclimation of cultures of various protozoa and bacteria
to increasingly high or low salinity levels has been
recorded frequently and has led to many studies investigating
the basis of salt tolerance.   Drought resistance, an
ecologically important trait, unfortunately has not received

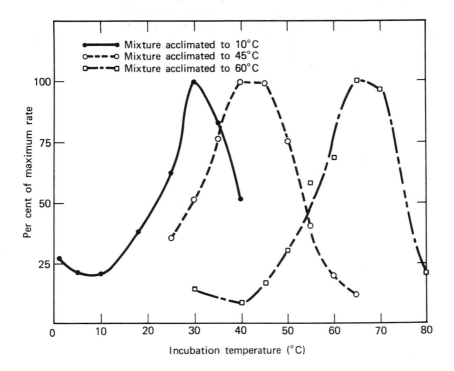

Fig. 4. *Effect of temperature on the rate of glucose
assimilation by microbial mixtures previously
acclimated to 10°C, 45°C and 60°C. Allen and
Brock (1968).*

the same attention. Among the toxic agents to which bacteria
or fungi have been shown to develop tolerance are mercury,
cadmium (Kondo *et al.*, 1974), copper (Kamalov and Ilyaletdinov,
1971), cyanide (Fry and Millar, 1972), and a large assortment
of fungicides, antibiotics, and other antibacterial and
antiprotozoan drugs. Some of the resistance mechanisms are of
purely theoretical interest, but many are not so; for example,
the adaptation of a plant-pathogenic fungus to a fungicide
is a serious problem, inasmuch as it leads to a loss of control
of the pathogen and the reappearance of the disease. Such
adaptations are evidently quite common in nature, and fungi
resistant to a fungicide to which the original pathogens were
sensitive have often been observed in field sampling programs.

Drug resistance among bacteria pathogenic to humans, similarly, is a widespread and commonly noted phenomenon. The appearance of penicillin-resistant strains of Staphylococcus in clinical specimens is depicted in Table 2.

TABLE 2

Penicillin-resistant staphylococci in hospital specimens[†]

Year	Strains resistant to Penicillin (%)
1946	5
1947–1948	18
1949	29
1950	44
1951	43
1952	31
1953	22

[†] From Zähner and Maas (1972).

The rise in relative abundance of tolerant strains during the first few years was linked with the increasingly frequent use of the antibiotic in chemotherapy, and the subsequent decline was attributed to the introduction of additional antibiotics into clinical practice beginning about 1950. The same type of change is evident in the data summarized in Figure 5, which shows a rise in the level of penicillin resistance among

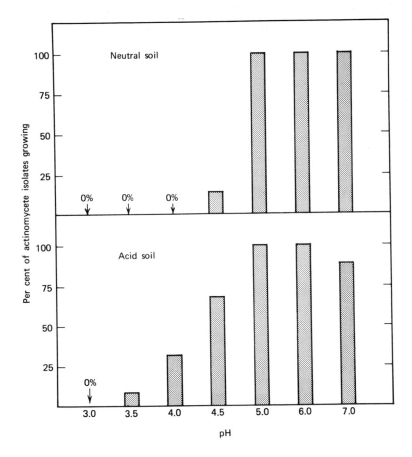

Fig. 5. *Rise in resistance to penicillin among isolates of*
Neisseria gonorrhoeae not eliminated by concentrations
of the antibiotic used in chemotherapy.
Thayer et al. (1965).

isolates of Neisseria gonorrhoeae. Apart from the practical
implications, these data demonstrate the same significant
ecological phenomenon: that continuing exposure of microorganisms to a harsh factor frequently results in the
appearance of strains able to withstand the deleterious
circumstance.

Of greater interest relative to Martian exploration is the fact that CO_2 apparently serves as an agent of natural selection. Although not a potent toxin, CO_2 is present in soil at partial pressures sufficiently high as to be inhibitory. That the gas may be a selective agent and CO_2 resistance may be an important fitness trait are suggested by the fact that at least some of the sensitive fungi are restricted to surface localities, whereas CO_2-resistant fungi, of the few organisms tested, exist at depths in soil where the CO_2 level is probably appreciable (Burges and Fenton, 1953). An apparent acclimation of soil microorganisms to CO_2 has also been reported (Stotzky and Goos, 1966).

Some of our work has been concerned with parasitism as a factor in selection and with those fitness traits that allow microorganisms to withstand attack by potential parasites. It is well known that soils contain fungi that produce dark-walled or melanized structures. These fungal structures are below the soil surface where the dark-walled coatings would not likely serve a protective role in preventing death of the organism by visible or ultraviolet light. The organisms forming these structures are frequently quite resistant to elimination, and persist for extended periods of time. The very same species frequently produce a mycelium which is rapidly eliminated, the resting stage, however, allowing the species to endure. The hyphae are rapidly parasitized by microogranisms in nature, and from such soils isolates have been obtained that produce extracellular enzymes capable of destroying the surface of the fungus and leading to the destruction of the underlying protoplast. The enzymes responsible for the digestion of the cell surface and consequent lysis of the hyphae are typically glucanases and chitinase. However, one fungus, Rhizoctonia solani, produces dark-walled hyphae, and these are characterized as being rich in melanin. The dark-walled hyphae, by contrast with the light-walled filaments of other organisms, are extremely persistent. Mutants of one melanin-producing fungus, Aspergillus nidulans, were obtained, and it was found that the melanin-containing walls were resistant to enzymatic digestion, whereas the essentially melanin-free walls of the mutant were quite readily degraded. On the basis of these findings, the otherwise resistant and dark resting stages were stripped of their melanin rinds, and the underlying materials were then readily digested by the lytic enzymes that attack the filaments. These findings demonstrate that melanin is a fitness trait among fungi, protecting them from elimination by lytic neighbors. Thus, there appears to be a favoring of organisms in soil which possess specialized structures that are not destroyed by potential parasites (Bloomfield and Alexander, 1967; Kuo and Alexander, 1967; Potgieter and Alexander, 1966).

Not all of the resistant fungi contain melanin; hence a study was made of two persistent fungi that were not readily attacked by lytic organisms, presumably because of a refractory component in their cell walls. The data from the investigations of Zygorhynchus and Mortierella suggested that a heteropolysaccharide is the likely reason for the longevity of these hyphae under natural conditions. Heteropolysaccharides probably owe their resistance to the many enzymes required to depolymerize a molecule having different monomers and different linkages, the array of enzymes required for a complex hydrolytic process probably requiring the coexistence, at a particular microsite, of a variety of different organisms. Hence, it is likely that the process would be reasonably slow (Alexander, 1973; Ballesta and Alexander, 1971; Pengra et al., 1969).

We have recently extended these observations to algae existing, not in soil, but rather in aquatic ecosystems. Some algae are quite resistant to elimination, and they persist in waters for many weeks with no evidence of significant parasitism. On the other hand, other species are readily attacked in fresh waters, and their cells are rapidly and appreciably degraded. Cell wall preparations of three of the susceptible algae, members of the genera Chlamydomonas, Cylindrospermum, and Ulothrix, were rapidly digested in enrichment cultures. The walls were largely solubilized by partially purified cellulase preparations in the case of Chlamydomonas and Ulothrix, and by a lysozyme preparation in the instance of Cylindrospermum. The same enzymes were without significant effect on three resistant algae, and the algal walls did not support any isolate in enrichment cultures. The addition of radioactively labeled walls to samples of natural ecosystems reinforced the view that the three species contained refractory components. Chemical studies were initiated to determine the biochemical basis of the resistance of the three long-lived algae. It was observed that Pediastrum walls have sporopollenin; Staurastrum contains a lignin-like compound; whereas Fischerella walls are rich in a heteropolysaccharide. Fractions obtained from the walls,

19

and containing these components, are likewise resistant to
attack by a mixed microbial inoculum as the test system for
biodegradation. Thus, it appears that fitness in these algae
is linked specifically with the presence in the walls of
unique structures--the sporopollenin, lignin-like compound,
and heteropolysaccharide being the presumed fitness traits
that allow the organisms to survive under the stress of attack
by parasites (Gunnison and Alexander, unpublished data).

Natural selection in sunlight-exposed environments, such
as fresh waters, leaf surfaces, and the atmosphere, favors
carotenoid-producing organisms. These pigments evidently
protect the organisms from being killed by the action of
sunlight (Table 3). The agent of selection is light, the
fitness trait is the carotenoid. The resistance of such
pigmented bacteria has been observed with Halobacterium
salinarium (Dundas and Larsen, 1962), Corynebacterium
poinsettiae (Kunisawa and Stanier, 1958), Rhodospirillum
rubrum (Cohen-Bazire and Stanier, 1958), and other microorganisms, illustrating again how a common stress, the lethal
action of sunlight, selects for organisms having a discrete
biochemical property.

Field observations reveal the frequency of melanized
fungal spores in light-saturated localities, as well as in
desert soils or the atmosphere. These colored structures
probably serve as effective light filters, shielding the
underlying protoplast from inactivation by incident radiation.
In line with this hypothesis are the frequent findings that
light induces fungi to produce melanin-rich or pigmented
spores.

The mechanisms of shielding from ultraviolet light have
been the subject of scrutiny, and the results are of interest
relative to regions where ultraviolet light intensity is high.
Thus, melanins appear to shield Nadsoniella nigra from ultraviolet light inactivation (Ruban and Lyakh, 1970), and
uncharacterized dark pigments or melanins seem to protect
fungi from lethal effects of gamma irradiation (Bondar et al.,
1971, Mirchink et al., 1968). A high rate of repair of DNA
may also be a means by which organisms survive in the presence
of gamma radiation (Davies and Sinskey, 1973).

Additional fitness traits of considerable ecological, not
to mention practical, importance include structural features
allowing for retention of microorganisms to surfaces in
habitats containing flowing water or even moving contents of

TABLE 3

Selective value of carotenoids in mixtures of <u>Halobacterium</u>
and a nonpigmented mutant †

Growth period	Ratio of red: colorless cells	
	Light[††]	Dark[††]
Inoculum	0.11	0.11
First[†††]	0.23	0.12
Second	0.86	0.12
Third	1.6	0.10
Fourth	6.8	0.12

†   From Dundas and Larsen (1962)

††   Cultures grown in light or dark

†††   End of first (etc.) serial subculture.

the alimentary tract; cellular components preventing aquatic
organisms from sinking in bodies of water; the mechanisms
associated with the development of virulence in pathogens in
the presence of their hosts; and the means by which animal
and plant parasites are protected from elimination by their
respective hosts (Alexander, 1971). Thus, one could easily
make a case for the relevancy of the ecologically irrelevant,
a case that one often had to plead in the past before

environmental sciences, even the basic facets, became of
interest to the general public.

EXOBIOLOGICAL ECOLOGY

A vast literature thus exists to confirm the view that
microorganisms can grow under a variety of extreme conditions.
Valentyne (1963) summarized the extremes of temperature,
oxidation-reduction potential, pH, hydrostatic pressure, and
salinity that appeared to serve as the limits for microbial
growth. Some of those limits can now be extended on the
basis of new findings, but more important are the additional
data pointing to the diversity of harsh circumstances which
microorganisms tolerate or to which they adapt.

What, then, are the environmental limits to life in this
solar system? Is it, in fact, possible to establish limits
of one or another extreme condition which will define the
tolerances of living things that have the only biochemistry
we recognize? On the basis of the available information, one
can propose ranges outside which no terrestrial organism will
exist, but the evolution of terrestrial species has proceeded
to only a certain extent in the direction of the extremes,
and one could easily conceive of the appearance of an
evolutionary sequence, on some other planet, which allowed
organisms to acclimate to conditions not tolerated by
inhabitants of this biosphere. The versatility of even the
microbial species we recognize suggests that representatives
of other biologies may have an enormous plasticity, one
allowing for the appearance of novel genotypes capable of
enduring the stresses on nearby planets. Indeed, several
investigators have shown the survival and even thriving of
earth organisms in media under circumstances presumably
similar to those on Mars or Venus. Thus, the adaptability of
living systems to the extreme conditions prevailing on other
planets of this or distant solar systems cannot yet be
delimited.

The tolerance ranges of microorganisms also have
implications for the problems of back contamination when
samples from Mars are returned to earth. At this time, it is
impossible to state with any degree of certainty whether
life exists on other planets of our solar system. However,
the capacity of at least some microorganisms, derived from
highly harsh environments on this planet, to grow at more
moderate conditions (Alexander, 1969) suggests that extreme
caution needs to be maintained--possibly by sterilizing all

returned samples or, with a higher degree of risk, by mounting
an extensive quarantine--lest a putative Martian be returned
to the earth to occupy some ecological niche in this biosphere
within its own range of tolerances.

REFERENCES

ALEXANDER, M. (1969) Nature (London) 222, 432.

ALEXANDER, M. (1971) "Microbial Ecology", Wiley, New York.

ALEXANDER, M. (1973) Biotechnol. Bioeng. 15, 611.

ALLEN, S.D., and BROCK, T.D. (1968) Ecology 49, 343.

BALLESTA, J.P.G., and ALEXANDER, M. (1971) J. Bacteriol. 106,
        938.

BLOOMFIELD, B.J., and ALEXANDER, M. (1967) J. Bacteriol. 93,
        1276.

BONDAR, A.I., ZHDANOVA, N.N., and POKHODENKO, V.D. (1971)
        Izv. Akad. Nauk SSSR, Ser. Biol. 80.

BRISOU, J., COURTOIS, D., and DENIS, F. (1974) Appl. Microbiol.
        27, 819.

BROCK, T.D. (1970) Annu. Rev. Ecol. System 1, 191.

BROCK, T.D., BROCK, H.L., BOTT, T.L., and EDWARDS, M.R. (1971)
        J. Bacteriol. 107, 303.

BURGES, A., and FENTON, E. (1953) Trans. Brit. Mycol. Soc. 36,
        104.

CHEN, M., and ALEXANDER, M. (1973) Soil Biol. Biochem. 5,
        213.

COHEN-BAZIRE, G., and STANIER, R.Y. (1958) Nature (London) 181,
        250.

COPELAND, J.J. (1936) Ann. N.Y. Acad. Sci. 36, 1.

CORKE, C.T., and CHASE, F.E. (1964) Soil Sci. Soc. Amer. Proc.
        28, 68.

DALLINGER, W.H. (1887) J. Roy. Micr. Soc. 7, 185.

DAVIES, R., and SINSKEY, A.J. (1973) J. Bacteriol. 113, 133.

DAVIS, J.S. (1968) J. Phycol. 4 (Suppl.), 6.

DOWBEN, R.M., and WEIDENMULLER, R. (1968) Biochim. Biophys. Acta 158, 255.

DUNDAS, I.D., and LARSEN, H. (1962) Arch. Mikrobiol. 44, 233.

FLOWERS, S., and EVANS, F.R. (1966) In "Salinity and Aridity" (H. Boyco, ed.) 367, Junk, The Hague.

FRY, W.E., and MILLAR, R.L. (1972) Arch. Biochem. Biophys. 151, 468.

GRIFFIN, J.L. (1972) Science 178, 869.

HOLM-HANSEN, O. (1967) "Environmental Requirements of Blue-Green Algae". Pacific Northwest Water Lab., Corvallis, Oreg.

JENKIN, P.M. (1936) Ann. Mag. Nat. Hist. 18, 133.

KAMALOV, M.R., and ILYALETDINOV, A.N. (1971) Vestn. Akad. Nauk Kaz. SSSR 27, (8) 44.

KAPLAN, I.R., and FRIEDMANN, A. (1970) Isr. J. Chem. 8, 513.

KONDO, I., ISHIKAWA, T., and NAKAHARA, H. (1974) J. Bacteriol. 117, 1.

KUNISAWA, R., and STANIER, R.Y. (1958) Arch. Mikrobiol. 31, 146.

KUO, M.J., and ALEXANDER, M. (1967) J. Bacteriol. 94, 624.

LACKEY, J.B. (1939) Public Health Reports 54, 740.

MEIKLEJOHN, J. (1957) J. Soil Sci. 8, 240.

MIRCHINK, T.G., KASHKINA, G.B., and ABATUROV, YU. D. (1968) Mikrobiologiya 37, 865.

MORITA, R.Y. (1967) Oceanogr. Mar. Biol. 5, 187.

MOSSER, J.L., MOSSER, A.G., and BROCK, T.D., (1973) Science 179, 1323.

ONISHI, H. (1963) Adv. Food Res. 12, 53.

PENGRA, R.M., COLE, M.A., and ALEXANDER, M. (1969)
        J. Bacteriol. 97, 1056.

PITT, J.I., and CHRISTIAN, J.H.B., (1968) Appl. Microbiol. 16, 1853.

POTGIETER, H.J., and ALEXANDER, M. (1966) J. Bacteriol. 91,
        1526.

RUBAN, E.L., and LYAKH, S.P. (1970) Izv. Akad. Nauk SSSR,
        Ser. Biol. 719.

RYTHER, J.H., and KRAMER, D.D., (1961) Ecology 42, 444.

SIEGEL, S.M., and ROBERTS, K. (1966) Proc. Nat. Acad. Sci.
        (U.S.) 56, 1505.

SKOWRONSKI, B., and STROBEL, G.A. (1969) Can. J. Microbiol.
        15, 93.

STARKEY, R.L., and WAKSMAN, S.A. (1943) J. Bacteriol. 45,
        509.

STEIN, J.R. (1968) J. Phycol. 4, (Suppl.), 3.

STOTZKY, G., and GOOS, R.D. (1966) Can. J. Microbiol. 12,
        849.

TANSEY, M.R., and BROCK, T.D. (1972) Proc. Nat. Acad. Sci.
        (U.S.) 69, 2426.

THAYER, J.D., SAMUELS, S.B., MARTIN, J.E., Jr., and LUCAS, J.B.
        (1965) In "Antimicrobial Agents and Chemotherapy--1964"
        (J.C. Sylvester, ed.) 433, American Society for
        Microbiology, Ann Arbor, Michigan.

TUTTLE, J.H., RANDLES, C.I., and DUGAN, P.R. (1968).
        J. Bacteriol. 95, 1495.

VALLENTYNE, J.R. (1963) Ann. N.Y. Acad. Sci. 108, 342.

VAN NIEL, C.B. (1955) J. Gen. Microbiol. 13, 201.

ZAHNER, H., and MAAS, W.K. (1972) "Biology of Antibiotics",
        Springer-Verlag, New York.

ANTARCTIC STUDIES

ELECTRON MICROSCOPY OF ANTARCTIC SOIL BACTERIA

Ian L. Uydess and Wolf V. Vishniac[†]

INTRODUCTION

Recent investigations in the arid, ice-free valleys of
South Victoria Land, Antarctica, have supported the view that
this region represents what is probably the most hostile
desert environment known on Earth. Precipitation, though not
infrequent, is sparse, and the relative humidity is only
about 10%. A number of investigators have reported that
these regions, except for several lakes and their surrounding
terrain, support a significantly lower number of bacteria per
gram of soil than do the soils of more temperate locales
(Horowitz et al., 1969; Benoit & Hall, 1970; Cameron et al.,
1970; Horowitz et al., 1972). In addition, it has been
reported, based upon direct attempts to cultivate microorganisms from these soils, that approximately 10% of the
locations examined were sterile or abiotic (Horowitz et al.,
1972). The inability to demonstrate microbial growth was
primarily attributed to the lack of water, which is one of
the most important limiting factors in these regions.

[†]

Deceased, December 10, 1973.

Vishniac and Mainzer (1973) explored many of these same areas during the austral summer of 1971-1972 and employed a relatively simple and straightforward technique to demonstrate microbial activity *in situ* in these regions. Sterile glass microscope slides were implanted into the soil at different sites in geographically distinct locations in the Wright Valley-Asgard Range region of South Victoria Land. Only actively multiplying bacteria that come in contact with the slides adhere to the glass substrate and form microcolonies of a few hundred to several thousand cells. The experimental stations selected represented a profile of the north slope of the Asgard Range from the floor of the Wright Valley, altitude 100 m, to the mountain crests, altitude 2200 m, where the diurnal soil temperature ranged from -5° C to +5° C. Soil samples were also collected from each site for cultivation in the laboratory. The authors reported that microcolonies of cells were observed on practically every slide that had been implanted into the soil. In addition, initial electron microscopic examination of soil films obtained by flotation (Dr. John Waid, Christchurch, New Zealand) disclosed the presence of a variety of micro-organisms including bacteria, algae (including diatoms), and, most important, large aggregates of bacteria which apparently were the result of cell growth in the locations from which the soil was originally collected (Figure 1). Furthermore, results on *in situ* labeled release experiments using 14 C-labeled carbon compounds, and experiments employing a light-scattering device originally developed for the Viking payload, supported the finding of microbial activity in these regions (Vishniac & Mainzer, 1973).

We now wish to report on the preliminary morphological observations we have made on several bacterial cultures isolated from three of these soil samples and to discuss the implications of certain of their ultrastructural traits concerning the ability of these organisms to survive in the arid soils of the Antarctic.

MATERIALS AND METHODS

Bacterial isolates were initially obtained from Antarctic soils by dispersing several grams of each soil sample upon the surfaces of growth substrates composed of either 2% Bacto-Agar (Difco Laboratories, Inc., Detroit, Michigan) or Sea Plaque Agarose (MCI Biomedical, Inc., Rockland, Maine) in a 1 ×, 10 ×, or 100 × dilution of Trypticase Soy Broth (General Biochemicals, Inc., Chagrin

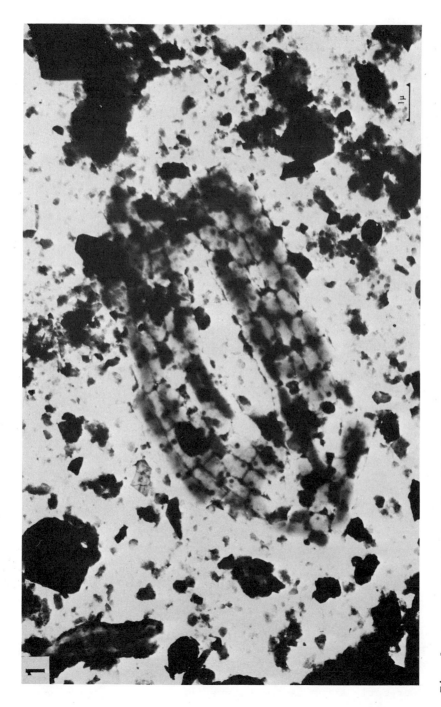

Fig. 1. *Large aggregate of cells in a negatively stained preparation of a soil film obtained by flotation (Dr. John Waid, Christchurch, New Zealand).*

1µ

1

Falls, Ohio) in isotonic, phosphate-buffered saline (PBS)
(Grand Island Biological Corp., Grand Island, New York).
These Antarctic soil "sprinkle plates" were then incubated
at 4° C for a period of up to six weeks. Colonies resulting
from the growth of organisms originally within the soil were
streaked out onto additional plates or slants to separate
mixed cultures. Subsequent "pure" cultures (selected from
single clones) were placed into separate tubes and assigned
an identification number.

Portions of each culture were removed from the surface of
the agar dishes or slants, suspended in 5 ml of isotonic PBS
and centrifuged at 8000 g for 10 min. The resultant cell
pellets were fixed in phosphate-buffered glutaraldehyde for
six to eight hours and in Dalton's Chrome Osmium (Dalton, 1955)
for twelve to twenty-four hours, and then postfixed in aqueous
uranyl acetate. Fixation was carried out at 4° C. The
samples were then dehydrated in a graded alcohol series and
embedded in Epon for electron microscopy. Thin sections were
prepared on an LKB Ultratome II (LKB Instruments, Inc.), using
glass knives. Sections were contrasted with uranyl acetate
and lead citrate, then photographs were taken on either a
Siemens Elminskop 1A or Philips 300 electron microscope at
80 or 100kV. A modified negative staining technique (the
OTPT procedure) was used to examine whole cell preparations
(Uydess, 1974).

RESULTS

Soil Sample No. 101

Negatively stained whole cell preparations of this
sample were initially observed to contain multisegmented,
single cells, 0.75 × > 2.0 μm, which exhibited numerous septa
and mesosome-like structures (Figure 2). The number of
segments present varied from 2 to 10 segments per organism,
with a majority of cells exhibiting between 4 and 7 segments.
Although the cells of this sample were usually observed to
have straight or slightly curved profiles, angular and club-
shaped individuals were not uncommon. However, no budding,
branching, or flagella were observed. Initial attempts to
visualize these cells in the light microscope using
conventional dyes were largely unsuccessful, as were attempts
to determine the Gram reaction.

Our early attempts to observe these cells in thin section

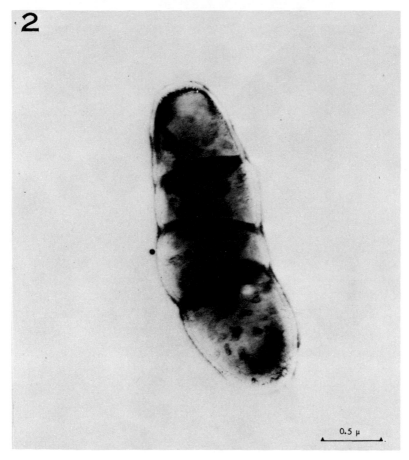

Fig. 2. Negatively stained whole cell from the 101 soil culture. Numerous septa and mesosome-like structures are evident in the multisegmented organism.

were similarly unsuccessful. Utilization of a routine fixation and embedding schedule resulted in an almost complete failure to preserve and infiltrate these organisms. Only after we employed a prolonged period of fixation and infiltration was a satisfactory result obtained. Examination of the subsequent preparations, however, yielded a surprising result. Although the multisegmented nature of these organisms was evident, no visible morphological continuum was apparent between adjacent segments, nor was there a visible cell wall (Figure 3 and 4). However, a limiting

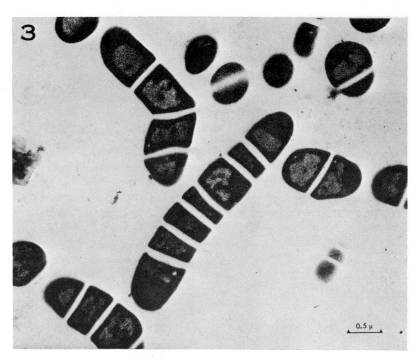

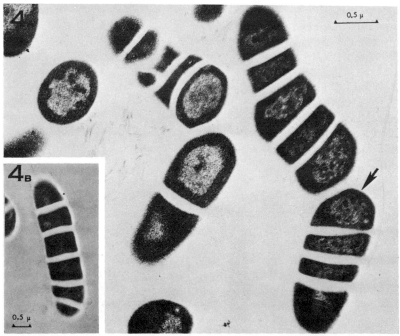

membrane was observed in a number of instances (arrow, Figure 4a), as were mesosome-like structures and membrane-bound vacuoles. The slight overexposure of a photographic plate during microscopy disclosed a uniform region of low electron density--a "halo"--between each cell segment, as well as around the entire periphery of each organism (Figure 4b). Thus, some substance was present in these areas which, although unstained by the reagents used in our procedure, exhibited a lower electron density than did the surrounding medium (Epon). The relationship of these cells to known families of bacteria, and the nature of the observed "halos" are as yet undetermined.

Negatively stained whole-cell preparations of a second isolate from the 101 soil sample contained cells of morphology similar to those previously described; they exhibited smooth surface topographies and numerous mesosome-like structures. The cells of this culture grew primarily as large aggregates, although single cells were occasionally observed (Figure 5 and 6). Unlike the previous culture, however, thin sections of this sample revealed cells with discrete cell walls (Figure 7a). In addition, it was observed that there was a heavy fibrous coat associated with the surface of these cells (Figure 7b). Subsequent determinative growth studies, based upon a number of growth parameters and their ability to form endospores, revealed that the cells of this sample were members of the family Bacillaceae.

Soil Sample No. 102

Thin section preparations of the major isolate from the 102 soil sample revealed that this culture was not pure, but was composed of at least two distinct groups of organisms. The first of these comprised a population of single cocci, 0.55 to 1.25 μm in diameter, which were limited by a multi-

---

Fig. 3 and 4. Thin sections of cells from the same 101 soil isolate as the cell in Figure 2. The multi-segmented nature of these organisms is evident, but there is no apparent morphological continuum between adjacent segments, nor is there a visible cell wall. Overexposure of a photographic plate during microscopy, however, did disclose the presence of a low density region between each segment and around the periphery of these cells (Figure 4b).

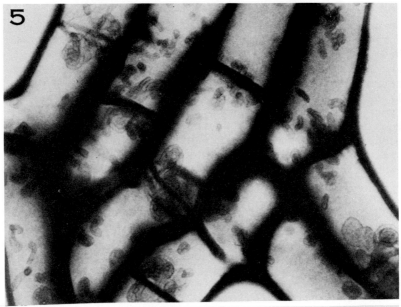

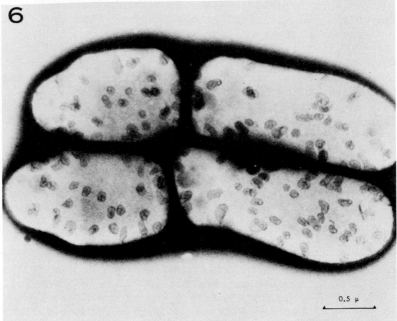

Fig. 5 and 6.   Negatively stained whole cell preparations of a
                second isolate from the 101 soil sample,
                exhibiting numerous mesosomes and smooth
                topographies similar to those of a number of
                known gram-positive bacteria.

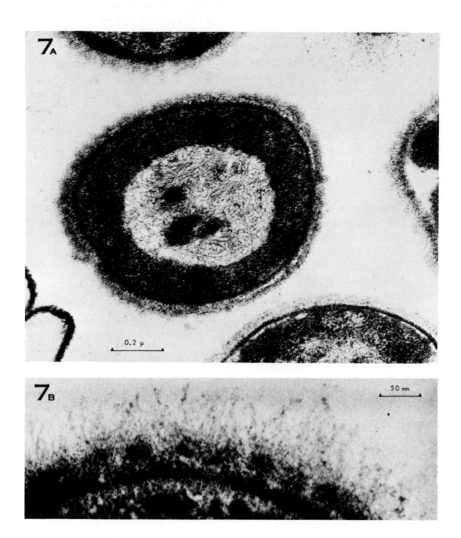

Fig. 7. *Thin sections of cells from a second isolate of the 101 soil sample, showing discrete cell walls and heavy fibrous coat.*

layered envelope and encapsulated within a dense, granular matrix (Figure 8). Morphologies similar to these have been reported previously by Bae and Casida (1973) for organisms isolated from the soils of more temperate regions. Aggregates of three to eight cells were also observed and were, on occasion, found to contain cells of apparently different morphologies (Figure 9 and 10). In several instances, cells exhibited membrane-bound cisternae which appeared to be continuous with a portion of the cell envelope (Figure 10a and 10b). The examination of whole cells of this sample by negative staining was unsuccessful, presumably due to the thickness of the capsule and the frequent occurrence of cell aggregates.

The second population of cells within this sample was found to contain two morphological types: (1) those cells which were observed to contain densely staining, ribosome-packed cytoplasms characteristic of known bacteria, and (2) those cells which exhibited significantly less dense cytoplasms, apparently devoid of discrete ribosomal structures (Figure 11a). In both instances the cells were bound by either a single or double "unit membrane" in the order of 60-80 Å. Furthermore, dividing cells within this sample apparently gave rise to progeny: (*i*) both of which contained densely staining, ribosome-packed cytoplasms (Figure 11b), (*ii*) neither of which contained densely staining cytoplasms (Figure 11c), or (*iii*) only one of which contained densely staining, ribosome-packed cytoplasm (Figure 11d). The basis of this observation is, as yet, unknown, however, artifacts resulting from the reaction of these cells (i.e., the lability of their components) to the fixation and embedding conditions used in these investigations cannot be overlooked.

Soil Sample No. 106

Attempts to observe the first isolate from the 106 sample by negative staining were again largely unsuccessful, owing to the extensive aggregation of cells. However, some detail in capsular structure was evident in relatively thin regions at the perimeter of cell aggregates (Figure 12). Thin sections of this sample revealed cells with coccoid to short rod morphologies, $0.6 \times 0.85 - 1.50$ μm, bound by a trilaminar envelope similar to that of many known gram-negative bacteria (Figure 13). The cytology of these organisms (i.e., the distribution of ribosomes and nucleic acids) was also similar to that of the known bacteria.

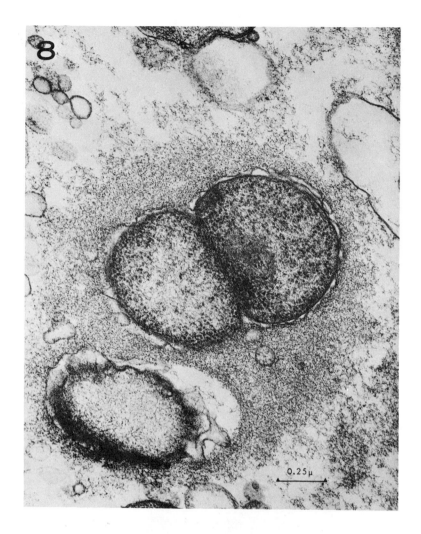

Fig. 8. *Thin section of a dividing cell from the 102 soil*
sample, exhibiting an extensive, fibrous capsule.

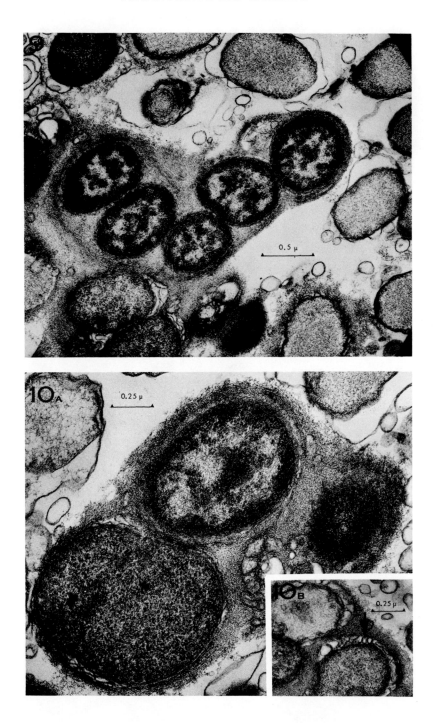

Fig. 9 and 10.   Encapsulated aggregate from the 102 isolate, exhibiting cells with multilayered envelopes and membrane-bound "cisternae" (arrow in Figure 10b).

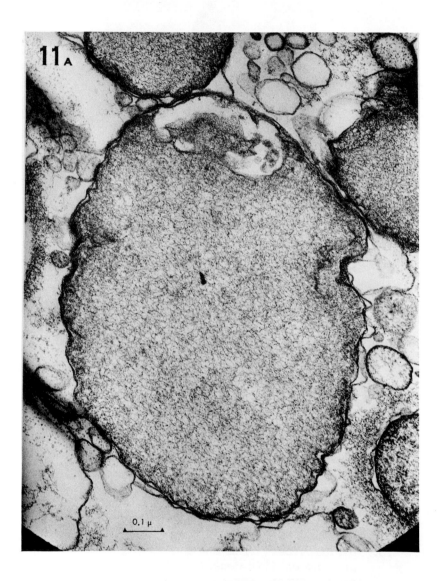

Fig. 11a.   Cell from the 102 sample which does not appear to have discrete ribosomes.

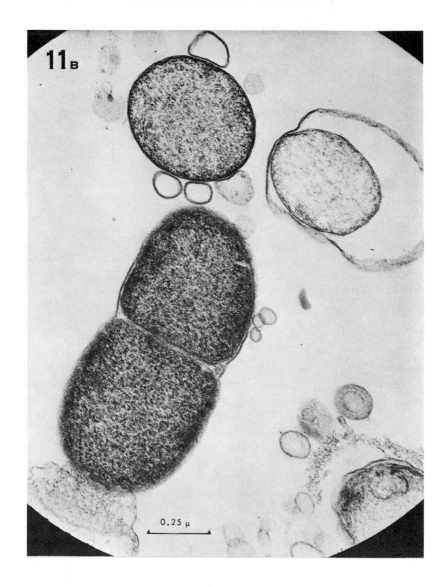

Fig. 11b. Dividing cell from the 102 sample, both of whose progeny exhibit densely staining, ribosome-packed cytoplasms.

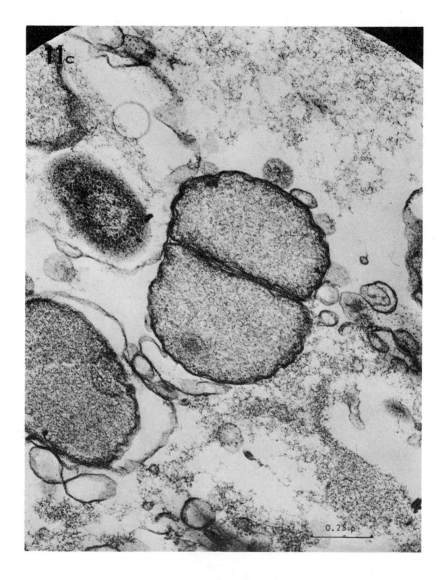

Fig. 11c. Dividing cell from the 102 sample, both of whose progeny lack densely staining cytoplasms.

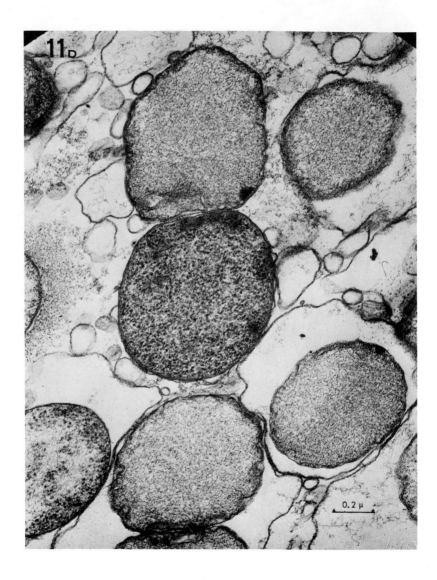

Fig. 11d. Dividing cell from the 102 isolate, only one of
whose progeny exhibits a densely staining,
ribosome-packed cytoplasm.

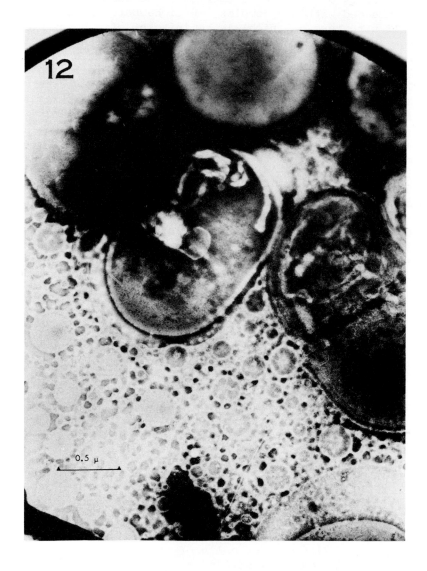

Fig. 12. *Negatively stained preparation of a cell aggregate from the 106 soil culture. Some detail in capsular morphology is evident in a thinly spread region at the perimeter of the aggregate.*

In addition, numerous membrane-bound vesicles were observed in a majority of cells. Furthermore, the cells of this sample were embedded within a complex, multichambered capsule, the overall structure of which was reminiscent of a honeycomb or sponge (Figure 13a). Each organism within the encapsulated aggregate appeared to be contained within a single compartment, or "cell". Membrane fragments and materials similar to nucleic acids and polysaccharides were also observed in localized regions within this structure, and may represent the partially degraded remains of dead cells (Figure 14).

Colonies of a second 106 isolate were observed to be pigmented (yellow-orange) when grown on TSB-Agar. Negative staining revealed single, monotrichously flagellated rods, 0.75×1.5 μm, and chains of four to six cells (Figure 15). In some instances, "minicells" and tubular bridges were observed between cells within such chains (Figure 15b and 15c). In addition, the surface topography of these organisms was similar in its convoluted appearance to that of a number of gram-negative bacteria (Uydess, 1974). Determinative growth studies disclosed that the cells of this sample were members of the family Pseudomonadaceae.

DISCUSSION

The extensive capsular material frequently observed in association with a majority of the Antarctic microorganisms may represent an ultrastructural trait which confers a selective survival advantage upon cells growing in dry soils. Given the sparsity of water characteristic of these regions, an extensive fibrous coat such as those described would represent a significant surface area over and within which water could be absorbed and held over extended periods of

---

Fig. 13. Thin sections of cells from the 106 soil sample, exhibiting laminar wall morphologies similar to those of a number of known gram-negative bacteria. In addition, the cells of this sample are observed to be embedded within an extensive capsule, the overall structure of which is reminiscent of a honeycomb or sponge. Each organism within the encapsulated population appears to be contained within a single chamber. A dividing cell, exhibiting mesosomes, is also shown (Figure 13b).

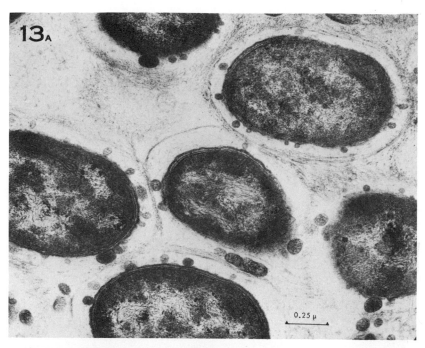

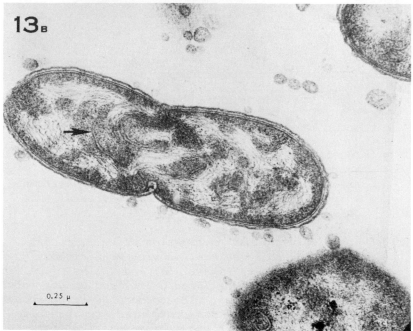

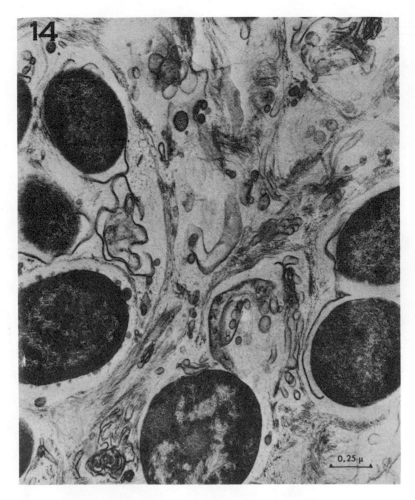

Fig. 14. *Localized region within an encapsulated 106 cell aggregate, containing cell wall and membrane fragments, as well as other cellular constituents, which may represent the disseminated remains of dead cells.*

Fig. 15. *Negatively stained preparation of cells from a second 106 soil isolate, showing monotrichously flagellated rods (Figure 15a) and chains of cells exhibiting minicells (Figure 15b) and unique tubular bridges (Figure 15c). The topography of these cells is quite similar in its convoluted appearance to those of a number of known gram-negative bacteria.*

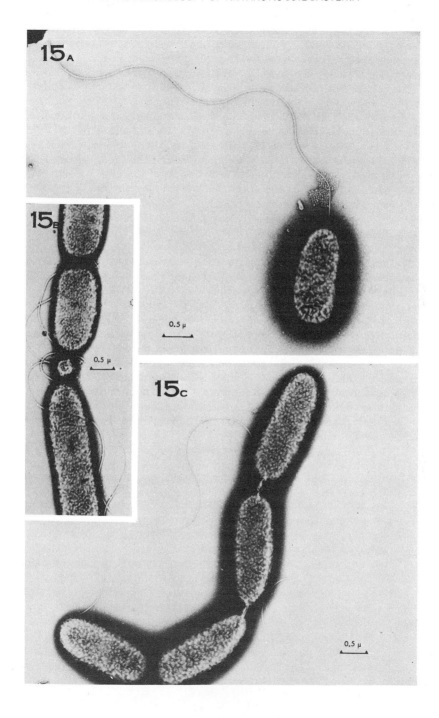

time. Bacteria possessing such capsules would be able to maintain themselves within a comparatively well-hydrated environment for intervals in excess of those normally afforded them by their natural soil environment. Furthermore, the aggregation of cells within such capsules would also confer a further survival advantage in allowing such cells to benefit from the metabolic activity of other individuals within that population (i.e., cross-feeding). In addition, the biochemical remains of dead cells within these encapsulated populations would serve to supply precursors of macromolecular synthesis otherwise not available to them in their Antarctic environment. The disseminated cellular components could then be effectively reutilized by viable cells in a continuing life-death cycle within the encapsulated microbial community (Figure 16).

Although relationships such as these are, at this time, only speculative and are based primarily upon the photographic data, they do afford a reasonable explanation for the ability of these cells to survive and grow in the arid soils of the Antarctic. Further detailed structural and biochemical studies are, however, needed in order to more accurately ascertain the validity of these proposed survival mechanisms. In addition, in order to more conclusively demonstrate that there is a viable (balanced) microbial community in the soils of these regions, it must be shown, *in situ*, that primary producers (e.g., autotrophic microorganisms) are present, and provide the organic materials depleted by the growth of other individuals in that community.

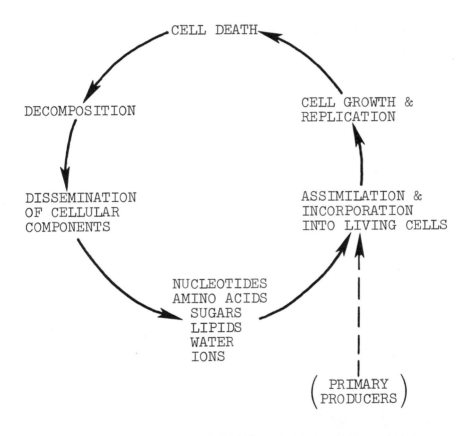

Fig. 16. *Diagrammatic representation of the proposed life-death cycle of Antarctic soil bacteria, demonstrating the possible reutilization of the disseminated cellular components of dead cells.*

ADDENDUM

Since this manuscript was originally prepared, a large number of microorganisms have been isolated from similar Antarctic soils collected at these and other sites during Dr. Vishniac's final journey to the "Dry Valleys" region of South Victoria Land, Antarctica, in 1974-1975. Included among these are 27 yeasts, 10 genera of filamentous fungi and 78 apparently dissimilar strains of bacteria (Hempfling, 1975). The isolates were obtained by a variety of cultivation techniques including sprinkle plates, shaker cultures, and a cascade enrichment procedure involving the percolation of soils in columns (Figure 16A) with a variety of hetero- and autotrophic (including photosynthetic) media. These results were contrasted to those employing more "traditional" methods of isolating soil microorganisms. All cultivation experiments were conducted at 4° or 10° C for a period of four to six weeks, after which the cultured soil samples were plated out onto minimal salts-agarose plates that were supplemented with the appropriate substrates, and then incubated for an additional four to eight weeks at 4° or 10° C.

Examination of shaker culture and percolated soils by transmission electron microscopy (TEM) (Uydess, 1975) revealed cells in association with soil particles (Figure 17 and 18). In addition, the direct examination by scanning electron microscopy (SEM), of E.M. grids that had been implanted--(*in situ*)--in the Antarctic soil for four to six weeks also revealed microorganisms, apparently cocci, adsorbed to the surfaces of individual soil particles (Figure 20 and 21).

Although we have not, as yet, been able to demonstrate the presence of strict chemoautotrophs (primary producers) in our soil cultivation experiments, Dr. E. Imre Friedman (Florida State University), to whom larger rock samples were sent, has recently informed us of the presence of endolithic algae within these samples.

On the basis of our experiments on isolation procedures we draw the conclusion that the more traditional methods of isolation of soil microorganisms are unsuited to reveal the existing microbial populations in the Dry Valley soils. Previous reports of sterile soils within these regions are probably due more to the utilization of inadequate isolation techniques than to the absence of living microorganisms.

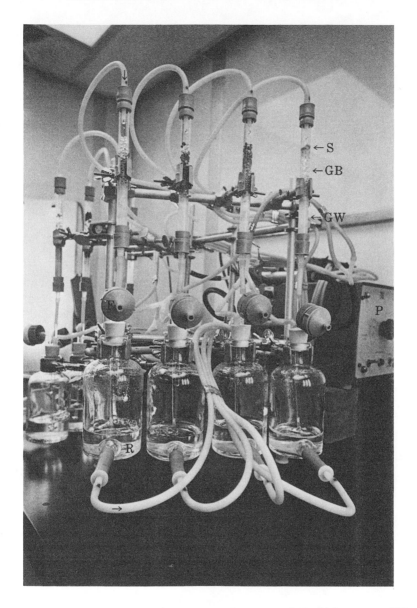

Fig. 16a. Soil percolation apparatus; peristaltic pump (P), media reservoir (R), soil (S), glass beads (GB), glass wool (GW), and exhaust filter (F).

Fig. 17 and 18. *Transmission electron micrograph of a negatively stained, percolated soil sample, showing cocci in association with soil particles.*

Fig. 19. *Scanning electron micrograph of a portion of the surface of a buried grid, showing numerous adsorbed soil particles.*

Fig. 20. *Scanning electron micrograph of the edge of a soil particle (dolerite), showing microcolonies of cells adsorbed to a brittle fracture surface.*

Fig. 21. *Scanning electron micrograph of cells on the surface of a soil particle.*

Fig. 22. *Scanning electron micrograph of a budding yeast cell isolated from a percolated soil sample prepared by critical point drying.*

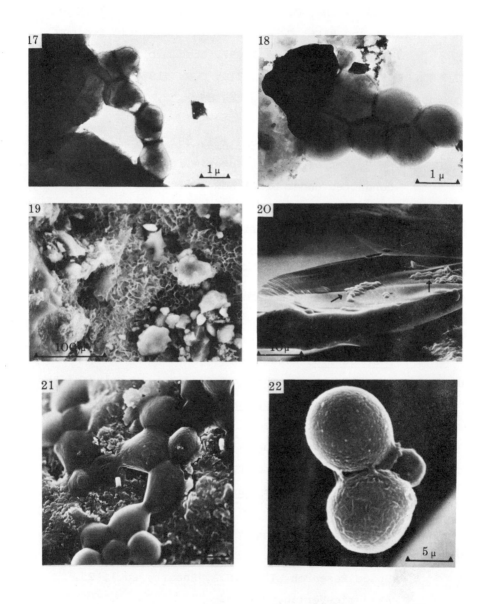

REFERENCES

BAE, H.C., and CASIDA, C.E. (1973) J. Bacteriol. 113, 1462.

BENOIT, R.E., and HALL, C.E. (1970) In "Antarctic Ecology"
(M.W. Holdgate, ed.) 697, Academic Press, New York.

CAMERON, R.E., KING, J., and DAVID, C.N. (1970) In "Antarctic
Ecology" (M.W. Holdgate, ed.), 2, 702, Academic
Press, New York.

DALTON, A.J. (1955) Anat. Record 121, 281.

HEMPFLING, W.P., et al. (1975) Interim report to NASA on work
conducted under Grant NGR 33 109 002; "Microbiological and Chemical studies of planetary soils".
Submitted to the Office of Planetary Biology,
National Aeronautics and Space Administration.

HOROWITZ, N.H., BAUMAN, A.J., CAMERON, R.E., GEIGER, P.J.,
HUBBARD, J.S., SHULMAN, G.P., SIMMONS, P.G., and
WESTBERG, K. (1969) Science 164, 1054.

HOROWITZ, N.H., CAMERON, R.E., and HUBBARD, J.S. (1972)
Science 176, 242.

UYDESS, I.L. (1974) Proc. Thirty-Second Ann. Meeting of the
Electron Microscopy Society of America
(C.J. Arceneaux, ed.) in press. Claitors Publishing
Company, Baton Rouge, La.

UYDESS, I.L., RICE, C.W., and HEMPFLING, W.P. (1975)
Proc. Thirty-Third Ann. Meeting of the Electron
Microscopy Society of America (C.J. Arceneaux, ed.)
Claitors Publishing Company, Baton Rouge, La.

VISHNIAC, W.V., and MAINZER, S.E. (1973) In "Life Sciences and
Space Research XI" (P.H.A. Sneath, ed.), 25,
Akademie-Verlag, Berlin.

ANTARCTIC MICROBIOLOGY—PREPARATION FOR MARS LIFE DETECTION, QUARANTINE, AND BACK CONTAMINATION

Roy E. Cameron, Richard C. Honour, and Frank A. Morelli

ANTARCTICA AS A MARS MODEL

A new phase of microbiological investigations began in the Antarctic in austral summer 1966. The purpose of these studies was similar to that of studies conducted in non-polar harsh environmental areas of five previous years, where hundreds of soil samples were collected (Cameron *et al.*, 1966) and microbial and ecological studies relative to Martian extraterrestrial experimentation and design (Cameron 1969, 1974) were performed both *in situ* and in the laboratory. The dry valleys and other exposed areas of the Antarctic are invaluable for study prior to searching for life on Mars, especially in relation to planetary quarantine and back contamination, because of characteristics of low temperatures and humidities; diurnal freeze-thaw cycles during a short productive season; low annual precipitation; high velocity desiccating winds; high sublimation, evaporation and transpiration rates; a low magnetic field and high radiation; absence of "higher" life forms, except for a few scattered members of the insecta, algae, lichens, and mosses; and a relatively low abundance of microorganisms when compared to temperate and arable areas.

During the short austral summer of four to six weeks, when there is a period of continuous daylight, there are diurnal fluctuations in environmental parameters within the dry valley region of South Victoria Land, not only because of the low angle of incident solar radiation, but also because of the interference of mountains surrounding the valleys, which have been deepened by glaciation. These environmental parameters have been recorded *in situ* for a number of sites, not only in the dry valleys, but also along the Transantarctic Mountain Range and as far south as Mount Howe (elev. 2800 m, 87° 21' S, 149° 18' W), the southernmost exposed mountain in the world, as well as actively volcanic Deception Island in the Antarctic Peninsula (Cameron, 1969, 1971, 1972, 1974; Cameron and Benoit, 1970; Cameron *et al.*, 1971a; Cameron *et al* ., 1970; Cameron *et al.*, 1971b).

Although the Antarctic has been contested as a relevant Mars model (Horowitz *et al.*, 1972; Vishniac and Mainzer, 1973), this region still contains the most appropriate study areas. Detailed information on the characteristics of Mars has been published previously, and can be used to compare with Antarctic data (Martin, 1972; Michaux, 1967; Sagan, 1972). The Antarctic obviously does not possess the same characteristics as Mars, but it has the Earth's harshest environment, one which is *most* analogous to Mars. It is far better to design and test Mars life-detection and other related experiments in the best available harsh terrestrial environment, than to study less harsh environments, or to perform *only* Mars-simulation tests based on the best available Mars environmental data.

These areas are useful for determining the abundance, distribution, and kinds of life forms that are closely linked to the environment. Such life forms exist as a few populations or as no detectable microorganisms at some sites as environmental factors become more limiting approaching the Pole (Jaffe *et al.*, 1974--see, especially, R.E. Cameron, Appendix A, "Sample Size for Biology" and Appendix F, "Temperature Requirements for Return of Sample"). A review of our Antarctic data for this eight-year period shows that approximately 10% of 10 g soil samples from the Antarctic may contain no culturable micro-organisms, as determined by ultimately sprinkling soil on agar plates of various media composition, and incubation conditions for extended time periods. This factor alone makes the Antarctic a valuable testing ground prior to determining the presence of any life forms with the use of the Viking in 1976 (Sagan, 1971; Young, 1973).

The comparative relationship of the Antarctic to Mars can be made clearer by considering it from the viewpoint of the biome concept. This concept can be applied to Mars, as well as to terrestrial regions, although it is not known at this time if Mars has any viable microorganisms. According to this concept, a biome is a geographical region consisting of a major biotic community, and composed of all the plants and animals of the community, including the successional stages of an area. The overall community also possesses certain similarities in physiognomy and in environmental conditions. For graphic purposes, a particular biome milieu can be plotted within the environmental parameter of annual precipitation versus temperature. For the Earth, these biomes have been previously ascribed to six regions or areas: (1) temperate and tropical desert, (2) grassland, (3) tropical forest, (4) deciduous forest, (5) coniferous forest, and (6) Arctic and alpine tundra (Figure 1).

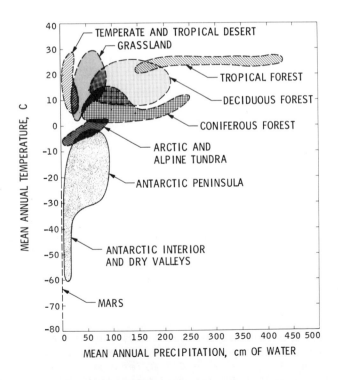

Fig. 1. Terrestrial and Martian biomes.

In addition, based on our own measurements plus known
Antarctic climatic data (Bentley *et al.*, 1964; Rubin, 1965,
1970; Rusin, 1964), the Antarctic biome has been plotted for
comparison with the more prominent and better known terrestrial
biomes and with the Mars biome. The Mars biome does not, of
course, have any measurable meteoric precipitation in the
centimeter range. On this basis alone it can be argued that
the Antarctic does not compare with Mars. However, the
Antarctic, for purposes of delimiting a biome milieu, can be
arealy divided into two geographic regions: (1) the
Antarctic Peninsula, which overlaps the Arctic in its more
favorable aspects of climate and vegetation, and approaches
some tundra conditions, and (2) a more distinct harsher and
barren geographical area primarily consisting of the
Antarctic dry valleys of South Victoria Land, approximately
78° S, 162° E, and south along the Transantarctic Mountain
Range to terminate at Mount Howe, 87° S. It can be shown for
the first time that, when compared with Mars as a prospective
biome, especially in regard to mean annual temperature of
the poles and equator (Michaux, 1967)†, the most Mars-like
geographical areas of the Antarctic are in the dry valleys
of South Victoria Land (but not restricted to them, since
there are other similar areas such as the Soviet base at
Mirny and the Pensacola Mountain Range) and the interior
(Figure 2). The High Arctic, especially in Northeastern
Greenland (Perryland), and lower latitude deserts, such as
the high elevations of the Atacama altoplano of Bolivia,
Chile and Peru, the Pamirs and Ladakh of Central Asia, are
among other close terrestrial approaches to a possible Mars
biome. The milieu of much of these areas is a physiognomic
dominance composed of microorganisms and cryptogams. In the
Antarctic, with increasing latitude, there are changes in
the density and kinds of life forms until only a few
populations of bacteria and cold-adapted (psychrophilic)
yeasts--less than 100 per gram of soil in some of the samples,
but not detectable in others--can be encountered in a 64 km^2
morainic area, 257 km from the geographical South Pole
(Cameron *et al.*, 1971a). For the Antarctic to be considered
as a model for possible life forms on Mars, it should be
understood that the Antarctic dates back to the Paleozoic,

---

† *Based on theoretical calculations, but not Mariner 9 data,
the mean annual temperature of the Martian equator may be
as low as -50° C and -100° C at the Poles (pers. commun.
C.M. Michaux, JPL Planetary Atmospheres Section).*

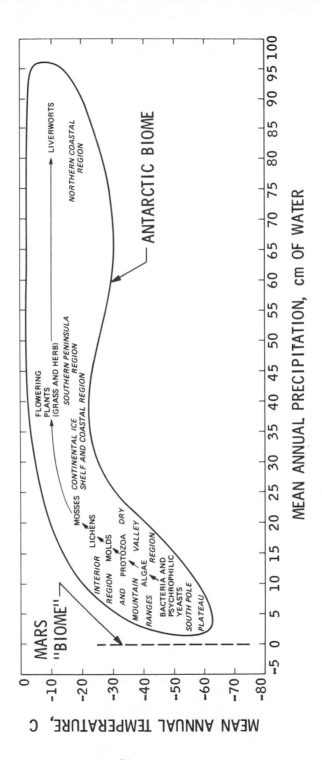

Fig. 2. *Antarctic and Martian biome milieu.*

61

with an ice age of more than 1,000,000 year B.P.   Therefore,
it may be quite unlikely that Mars life forms will exceed
the complexity of that of a few scattered lichens.

In consideration of Antarctic life forms, the population
density, abundance, diversity, complexity, and size varies
with the degree of favorability of environmental factors
(Figure 3).   These consist of major factors of   (1) climate
and microclimate--especially in terms of the period of
interrupted temperatures above freezing,   (2) topography--
especially a northern exposure near a source of available

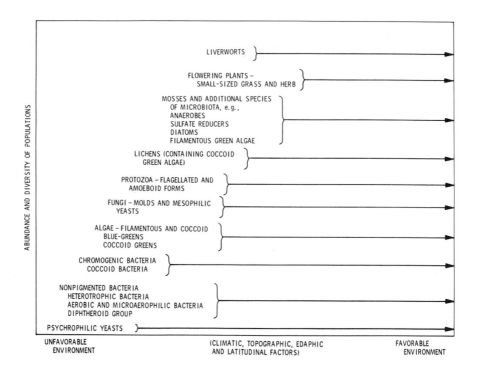

*Fig. 3.   Variability of population density and diversity with
variability of ecological factors in undisturbed
Antarctic soils.*

(liquid) moisture, (3) edaphology--soil in relation to its properties favorable for growth of microorganisms and cryptogams, and (4) lower Antarctic latitudes and elevations. An established plant community, such as that observed for the algae in the dry valleys, is generally in the proximity of glacial runoff, meltwater pools, and streams; and it is atypical in terms of the climatic environmental parameters of mean annual precipitation and moisture; it really should be considered existing in a more favorable microenvironment. The algae depend upon meltwater as a source of available water during the short austral summer which is interrupted by diurnal freeze-thaw cycles; they are not active in the most arid regions of the dry valleys. Their occasional presence in samples procured from dry areas (Vishniac and Mainzer, 1973) should not be misconstrued as current and active endemicity, since the Antarctic is a natural "freeze-dryer" and the winds, along with birds and man, are responsible for widely distributing the biomass and windborne trash from more productive areas into unreceptive areas. This is especially true for diatoms, which are found in more favorable habitats where moisture is available. It should also be considered, as has been found by various investigators (Benoit and Hall, 1970; Cameron *et al.*, 1970) that there is a positive correlation between the time visible growth of microorganisms occurs on an isolation medium, and the time interval since conditions were last favorable for growth at a given site. In locations where growth conditions had not been favorable for many years, only a few psychrophiles were detected (Benoit and Hall, 1970). In such cases, the lag period necessary for activation from dormancy may approach four weeks to three months (Benoit and Hall, 1970). For the Antarctic algae, more than one year of incubation of moist soil may be necessary before visible growth of pinpoint colonies is observed (Cameron 1972b; Cameron *et al.*, 1971a). Unfortunately, for Antarctic microorganisms "heat shock" is a risk, because it may kill the microorganisms rather than break the dormancy period.

The algae not only are inactive in the most arid areas of the dry valleys, but are also not able to withstand the rigorous cold, arid, and shorter productive period of the polar plateau in the exposed area closest to the South Pole. These algae have been found only as far south as a frozen pond in the La Gorce Mountains (86° 45' S, 146° 0' W) (Cameron, 1972b). The protozoa and molds that are dependent upon the algae as a substrate, and for nutritional and protective purposes, do not actively exist separate from the algae, except in areas that have been organically contaminated. Scattered lichens, whether on the floors and at the mouths of the dry valleys, or on the sides of the lower elevated valley walls and higher elevated hanging valleys (e.g., Wheeler Valley, King-David, or Odin Valley), can only inhabit sites or strips between the 80% relative humidity line and the snowline (Wilson, 1970). The farthest south that lichens have been found was at Moraine Canyon in the Queen Maud Mountains (86° 09' S, 157° 30' W), where they had a protected, yet favorable microhabitat with respect to snow-melt, direction of exposure, and microtopography.

ANTARCTIC PERTURBATIONS

A word of caution needs to be expressed in regard to current investigations of Antarctic microbial ecology. The Antarctic, especially in the dry valley region of South Victoria Land, has been altered in recent years, although not visibly so, in regard to its microbiota, through the incursions of man (Cameron, 1972c). These alterations include the establishment of semipermanent field camps, such as the Lake Bonney Hut, Lake Vanda Station for "winter-over" parties, the use of motorized field vehicles and helicopters, and repositories of materials and supplies at former field camps. As shown by the study of many Antarctic areas, the Antarctic ecosystem, especially in the driest areas, is a very simple, and also a very fragile ecosystem. It can, and has been, easily impacted or perturbed, altered and/or destroyed, especially by the activities of man, but also by vectors such as the camp-following skuas. Although a few formerly unperturbed areas contained only a single population of bacteria or yeasts, or no detectable microorganisms, there are increasing numbers of bacteria and molds that were not detected formerly in the Antarctic.

The bacteria in unperturbed or pristine Antarctic soil ecosystems are not sporeformers, but with the spread of man's activities and incursions into formerly pristine areas, sporeforming Bacillus spp. are becoming more prevalent in the Antarctic. These microorganisms were not found in newly investigated areas in the Pensacola Mountain Range. However, the abundance or absence, and kinds of microorganisms, resembled those seen in earlier investigations of the dry valley region of South Victoria Land (Cameron and Ford, 1974). The absence or scarcity of sporeformers in relatively undisturbed areas has been confirmed by others (Vishniac,

1973), and it also was noted, as we have observed, in polar as well as in very arid nonpolar areas, that the limited availability of moisture does not allow enough time for spore formation and germination. Only the psychrophilic yeasts may have a relatively rapid growth rate, and they may be killed at temperatures above 15° to 20° C.

Man's impact can be so severe as to wipe out the indigenous Antarctic species, or to completely alter the indigenous community. Some of these observations are summarized in Tables 1-3. As far as the bacteria are concerned, many of the colonies which have been isolated from Antarctic soil, water, and air do not appear to be indigenous, as shown by a comparison with species previously recorded from various non-Antarctic habitats (Cameron *et al.*, 1972a), and except for the sporeforming Bacillus spp., many could be classified as coryneform "antifreeze bacteria" because they probably have protective lipid cell walls and/or contents. As based upon original source of isolation, metabolism, (especially nitrification and denitrification), optimum growth temperatures, and salt tolerance, it appears that there are few indigenous Antarctic bacteria (Table 4).

There has also been an increase in the number and kinds of fungi in areas formerly lacking in this group of micro-organisms. Penicillium spp. are becoming much more prevalent. During the tenure of the Dry Valley Drilling Project (DVDP) team in the Antarctic last austral summer, a Neurospora sp. was introduced, probably in contaminated food, at the last drill site of the season (New Harbor, Taylor Valley). This same mold later became an obnoxious contaminant at the Eklund Biological Center at McMurdo Station (Cameron *et al.*, 1974a,b). The yeast Cryptococcus albidus was one of the most prevalent psychrophilic yeasts, but this same species also is found in many temperate areas, as well as in frozen foods. A list of molds and yeasts recovered during our studies in the Antarctic is given in Tables 5 and 6.

MARS LIFE DETECTION AND QUARANTINE PROBLEMS

At the present time it is still unknown as to whether or not there are viable life forms on Mars. Even though the 1976 Viking Lander may not be able to detect life, it is still possible that Martian life forms, at least on the microbial level, may exist. The Viking Lander is recognizably

TABLE 1

Comparison of Populations in Disturbed versus Undisturbed
Antarctic Soils

Microbiota enhanced by site disruption

  Heterotrophic, aerobic bacteria

  Microaerophilic bacteria

  Heterotrophic, anaerobic bacteria

  Chromogenic bacteria,
    yellow and orange colonies

  Ammonifiers

  Hydrocarbon decomposers

  Cellulose decomposers

  Denitrifiers

  Sulfate reducers

  Sporeforming bacteria

  Bacteria capable of surviving
    37° C to 55° C

  Bacteria with shorter growth period

  Molds and yeasts

TABLE 2

Comparison of Populations in Disturbed versus Undisturbed
Antarctic Soils

Microbiota eliminated by site disruption

    SOME BACTERIAL GROUPS, e.g.,

        Fastidious reds

        Diphtheroid colonies

        Nitrifiers

        Nitrogen fixers

    SOME YEASTS

    ALGAE

    SOME PROTOZOA

    LICHENS

    MOSSES

TABLE 3

Comparison of Populations in Disturbed versus Undisturbed
Antarctic Soils

Microbiota shifted from natural site sequence

Nonpigmented bacteria

Chromogenic bacteria

Streptomyces group

Molds and yeasts

Protozoa

Heterotrophic,
  anaerobic bacteria

Sulfate reducers

TABLE 4

Antarctic Bacteria

Actinomycetales

Mycobacteriaceae

Mycobacterium sp.

Mycococcus albus[a]

Mycococcus ruber[a]

Mycococcus sp.

Actinomycetaceae

Streptomyces albus[a]

Streptomyces exfoliatus[b]

Streptomyces longisporoflavus[a]

Streptomyces parvus[b]

Streptomyces sp.

Nocardia albicans[a]

Nocardia flava[a]

Eubacteriales

Achromobacteriaceae

Achromobacter butyri[b]

Achromobacter liquefaciens[b]

Achromobacter parvulus[b]

Achromobacter stenohalis[b]

Achromobacter superficialis[b]

Achromobacter xerosis[b]

Flavobacterium diffusum[b]

Flavobacterium solare[b]

Bacillaceae

Bacillus cereus[b]

Bacillus circulans[b]

Bacillus coagulans[b]

Bacillus firmus[b]

Bacillus lentus[b]

Bacillus megaterium[b]

Bacillus pergrinosus[b]

Bacillus pumilis[b]

Bacillus subtilis[b]

Bacillus sp.[b]

Brevibacteriaceae

Brevibacterium ammoniagenes[b]

Brevibacterium fulvum[b]

Brevibacterium incertum[b]

Brevibacterium imperiale[b]

Brevibacterium maris[b]

Brevibacterium sulfureum[b]

Brevibacterium tegumenticola[b]

Kurthia bessonii[b]

Kurthia variabilis[b]

Corynebacteriaceae

Arthrobacter citreus[a]

Arthrobacter globiformis[b]

Arthrobacter simplex[a]

Arthrobacter tumescens[a]

Arthrobacter ureafaciens[b]

Arthrobacter sp.

Corynebacterium bovis[b]

Corynebacterium capitovale[b]

Corynebacterium equi[b]

Corynebacterium hoagii[b]

Corynebacterium hypertrophicans[b]

Corynebacterium peregrinum[b]

Corynebacterium pseudodiphthericum[b]

Corynebacterium rathayi[b]

Corynebacterium sepedonicum[b]

Corynebacterium striatum[b]

Micrococcaceae

Micrococcus albus var. albidus[b]

Micrococcus candidus[b]

Micrococcus conglomeratus[b]

Micrococcus flavus[b]

Micrococcus freudenreichii[b]

Micrococcus rubens[b]

Micrococcus ureae[b]

Sarcina flava[b]

Sarcina frigia[a]

Sarcina sp.

Staphylococcus epidermidis[b]

Staphylococcus pyogenes[b]

Pseudomonadales

Pseudomonadaceae

Pseudomonas fragi[b]

---

[a] *Possible indigenous species.*

[b] *Possible introduced species.*

71

TABLE 5

Antarctic Fungi

MOLDS

Phycomycetes

Mucor jansseni

Rhizopus nodosus

Ascomycetes

Arachniotus citrinus

Eurotium (Aspergillus) repens

Deuteromycetes

Alternaria tenuis

Aspergillus versicolor

Aureobasidium

Chrysosporium merdarium

Chrysosporium pannorum

Chrysosporium verrucosum

Cladosporium sphaerospermum

Dendryphiella salina

Fusarium sp.

Gliocladium sp.

Helminthosporium anomalum

Monodictys austrina

Paecilomytes sp.

Penicillium adametzi

Penicillium canescens

Penicillium chrysogenum

Penicillium citreo-viride

Penicillium corylophilum

Phialophora dermatitidis

Phialophora fastigiata

Phialophora gougerotii

Phialophora lagerbergii

Phoma hibernica

Sepedonium chrysospermum

Trichoderma harzianum

Tritirachium album

Tritirachium roseum

TABLE 6

Antarctic Fungi

YEASTS

Aureobasidium pullulans

Candida albicans

Cryptococcus albidus[a]

Cryptococcus albidus var. diffluens[a]

Cryptococcus lateolus

Pullularia sp.

Rhodotorula graminis

Rhodotorula minuta

Rhodotorula graminis

Sporobolomyces salmonicolor

Torulopsis sp.

---

[a] *Includes obligate psychrophils.*

limited not only by its few methods of life detection, but
also by the following:

(1) It will not be able to land at high latitudes
(> 70°), in the vicinity of visible permafrost
where the possibility of life may be much higher.

(2) It probably will not land at possible, but widely
scattered, favorable microhabitats.

(3) It does not have the technical capability of
selectivity and maneuverability in regard to
landing sites.

(4) It is designed to only scratch the surface and does
not have the capability to obtain subsurface
samples, which, in consideration of Mars's
environmental conditions may contain entrapped
water in the form of permafrost.

The probability factor for possible life on Mars, and
transport of viable life forms to or from Mars, is still, at
best, an educated guess rendered by experts, based on
(1) best available knowledge of Mars surface and atmospheric
features, (2) simulated laboratory tests for survivability
and adaptability of microorganisms from various terrestrial
sources, especially the familiar contaminants of clean rooms,
and (3) the occurrence, distribution, abundance, and kinds
of naturally occurring terrestrial environments, where
environmental conditions are limiting for life. In regard to
this latter factor, the research has been limited by the
number of qualified workers who are willing to undertake the
rigors and hardship of field research under adverse conditions,
or who lack the flexibility of modifying, adapting and
applying laboratory methodology to field conditions. This
has consequently resulted in a lack of needed analogical
information applicable to possible Mars life detection and
related quarantine problems, including (1) possible transport of viable terrestrial microorganisms to Mars, (2)
possible subsequent survivability and/or adaptability, (3)
return of viable indigenous Martian microorganisms to Earth,
and (4) best possible methods of containment, survivability,
and/or adaptation and proliferation of any likely indigenous
Martian microlife forms in the seven major terrestrial biomes,
that is, temperate and tropical desert, grassland, tropical
forest, deciduous forest, coniferous forest, Arctic and
alpine tundra, and Antarctic regions.

Although extreme precautions and procedures were first taken for quarantine and handling of returned Lunar samples, it should be recognized that these elaborate, time-consuming, and expensive measures may not be necessary for return of unsterilized Mars samples. Although it is realized that there are two "camps of opinion" in this matter, (Alexander, 1972; Horowitz *et al.*, 1972; Vishniac and Mainzer, 1973), it may be considered that the chance that the Viking Lander will find life on Mars, based especially on the Lander's technological and engineering limitations, is not likely. That the sample may be returned to Earth with its possible indigenous life forms maintained in viable condition, is another hazard to consider (Jaffe *et al.*, 1974); that these same life forms could even break through a containment barrier, and finally, that they would endanger, adversely affect, or take over and eliminate Earth populations may be quite remote. However, on the basis of Antarctic microbial ecology, the impact of man's activities and the importation and spread of exogenous or contaminating species from other biomes is becoming increasingly noticeable. Imported mesophilic and higher temperature strains are surviving and spreading in Antarctic ecosystems. Contrariwise, Antarctic species are difficult to maintain, let alone to proliferate under optimal, controlled conditions in the laboratory. Even room temperature storage can endanger their survival. It can therefore be concluded that the geographically isolated and formerly pristine Antarctic continent will continue to suffer the impact of introduced competitors. It may well be that the camp of advisers who envision the escape, endangerment, or take-over and elimination of any or all Earth forms, are merely indulging in the realm of science fiction, but if organisms do escape, they will most likely survive and/or adapt in those terrestrial areas which most closely approach that of a Mars biome. This could be quite important, not only from the viewpoint of possible Martian life forms, but also in regard to possible endangerment of terrestrial life forms in case a returned Martian life form was not contained, and escaped into the Earth environment. As shown in Figures 1-3, the most likely sequence of endangerment from escaped Mars microorganisms would be (1) the Antarctic interior, (2) the Antarctic dry valleys, (3) the Antarctic coastal areas and peninsula, (4) the high Arctic and tundra, and (5) the altoplano and other geographical areas approaching high latitudinal conditions.

A recent development that should be considered in regard
to life detection on Mars, quarantine, and back contamination
is, it appears that viable fossil microorganisms have
recently been revived from subsurface (approaching 400 m
depth) Antarctic permafrost dated at 10,000 to > 1,000,000
years B.P. (Cameron and Morelli, 1974). These microorganisms
were not sporeformers, but were coryneform "antifreeze
bacteria". The presence of other microorganisms, including
diatoms and microfaunal cysts, indicated a microbial community
not dissimilar from that found in more favorable Antarctic
microenvironments today, or in favorable microenvironments of
arid temperate regions. Long-term survival of microorganisms
in Martian permafrost has been postulated previously
(Sagan, 1971, 1973; Milton, 1974); the millenial survival of
bacteria within ice plates of shallow (40 cm) Antarctic
permafrost also has been postulated (Vishniac, 1973). The
necessity for longer term incubations, lower viability, and
the different nature of the microbial community in subsurface
samples has been shown by our previous work. Although the
1976 Viking Lander may not find viable microorganisms with
its present methodology of sampling and analysis, it should
be stressed that negative results do not rule out the strong
possibility of viable life forms in the permafrost.
Proposed post-Viking studies (Young, 1974) should therefore
consider this intriguing possibility.

Another possibility is that Martian microorganisms may
exist in very special microenvironments, such as the salt
ponds which are found in the Antarctic. Don Juan Pond, a
pond of saturated calcium chloride hexahydrate (antarcticite)
is in equilibrium with the air at 39.8% R.H. Other ponds, such
as Don Quixote Pond show similar conditions, and it may be
that microorganisms are adapting to this extreme environment
(Cameron *et al.*, 1972b). It has been hypothesized that
natural heating in the lower strata of salt ponds is the
result of the combination of refraction and absorption of
solar energy in the lower saline layers of these ponds
(Hudec and Sonnenfeld, 1974). Most of the recoverable
Antarctic microorganisms are mesophiles; it may be that
long-term warming of these ponds may have provided a special
habitat for the adaptation of halophile/mesothermophiles.
Cores drilled from beneath these ponds may yield recoverable
microorganisms showing a special case of survival as a
result of reduced competition in a specific physical and
chemical environment; they would reflect the microbiota of
the warmed salt pond and not of the surrounding environment.
The possibility of finding such microorganisms in the

Antarctic, with reference to Martian life forms, is indeed exciting, especially in view of the history of probable water degradation on the surface of Mars (Milton, 1973, 1974; Sagan, 1973).

MARS SIMULATION AND CONTAINMENT STUDIES

Terrestrial microorganisms from more favorable environmental areas, including the Arctic tundra, are probably much more competitive and less vulnerable than Antarctic microorgamisms, although introduced species, after they have survived and adapted to Antarctic conditions, may be more competitive if returned to their former environments. There are many laboratory and field tests that could be performed along this line, and the understanding of survival and/or growth of Antarctic microbiota under the severe, moisture-limiting freeze-thaw conditions of the short-term austral summer still needs further study. Based upon Antarctic studies, it may be more of a challenge to maintain a Mars microorganism alive than to be concerned about its containment and possible proliferation if it escapes into the destructive microbe-eating terrestrial environment, especially at lower latitudes, where the samples will most likely be received and studied.

An elaborate biological barrier system may not be necessary. A "heat shield", for example, a "Sauna bath", may be all that is necessary. This Sauna bath would be an enclosure or entrance way to the actual repository where the Mars samples are maintained, or worked on, under conditions more favorable for maintaining the integrity and biological quality of the samples. However, a second circumferential cold room or entrance could precede the entrance to the Sauna bath. This latter room could contain exposed culture media of various composition at below freezing temperatures to assist in determining if any Mars microorganisms had survived the heat shield. The cold room could be equipped with a germicidal (UV) light system to further inhibit or destroy any possible escapees.

Simulation tests for survival and/or adaptation of Martian microbial life forms could be based on variations of environmental parameters related to the three factors proposed for the barrier shield as just outlined, for example, (1) increasing levels of temperature, (2) increasing levels of moisture (room humidity), (3) combinations of (1) and (2), (4) subsequent exposure to cold, (5) exposure to UV levels, and (6) combinations of cold and UV, including a variety of culture media and entrapping mechanisms. A wide variety of test materials are available outside the Antarctic continent, including soil, ice and snow stored in an "Antarctic Simulator", as well as soils from many other biomes of the world, including tundra, high mountains, temperate and tropical deserts, which were collected for NASA and NSF programs over a thirteen-year period. Non-Antarctic soils of known properties, both biotic and abiotic, could be mixed with viable Antarctic soils, or inoculated with known kinds and concentrations of single or mixed Antarctic microbial isolants to determine rates of survival and/or adaptation, and possible proliferation of microorganisms. Such tests will give a much better understanding of possible Mars life detection and quarantine problems, and help to formulate a much better Planetary Quarantine probability factor.

ACKNOWLEDGEMENT AND EULOGY

Appreciation is expressed to the many who have encouraged and made possible the past eight years of study of Antarctic microbiology and ecology. During the tenure of most of this study, funds were supplied primarily by NSF Grant C-585 and NASA Contract NAS7-100 to the California Institute of Technology, Jet Propulsion Laboratory, with logistic support by the U.S. Navy and VXE-6 Helicopter Squadron. For the last austral summer in the Antarctic, funds were provided through NSF Grant GV-40602 to the Darwin Research Institute.

Appreciation also is expressed to Dr. Richard S. Young, Office of Space Science and Applications, NASA Headquarters, for his recommendation that this paper be presented at this Symposium in honor of Dr. Vishniac, and especially in recognition of his personal interest in the Antarctic, relative to Mars life detection and contamination.

Frank Morelli, a staff member of DRI, gave the eulogy for Dr. Vishniac at McMurdo Station and assisted with other details following his death. Mr. Morelli was later joined by the senior author in a visit to Dr. Vishniac's campsite and place of death in upper Wright Valley. A visit to these sites shows the Antarctic in its stark reality, hypnotically majestic, yet treacherous and unforgiving, even to exacting the supreme penalty for a single misstep!

The authors express their admiration for Dr. Vishniac's active interest in the Antarctic and his fortitude as a member of the Viking Committee to "see it for myself" though he lost his life in the process. His great fascination for the pristine beauty of the Antarctic, as well as his determination to study it for its scientific value *per se*, are shared by those of us who have returned again and again to learn the secrets of this "Great White Desert". May the future explorers of both the Antarctic and Mars pause for a moment as they consider the dedication and sacrifice of those who have preceded them.

REFERENCES

ALEXANDER, M. (1972)   In "Theory and Experiment in Exobiology" (A.W. Schwartz, ed.), 2, 123. Wolters-Noordhoff, Groningen.

BENOIT, R.E., and HALL, C.L., Jr. (1970)   In "Antarctic Ecology" (M.W. Holdgate, ed.), 2, 697. Academic Press, New York.

BENTLEY, C.R., CAMERON, R.L., BULL, C., KOJIMA, K., and GOW, A.J. (1964)   In "Antarctic Map Folio Series" (V.C. Bushnell, ed.), Folio 2. American Geographical Society, New York.

CAMERON, R.E. (1969)   In "Arid Lands in Perspective" (W.G. McGinnies and B.J. Goldman, eds.) 169. AAAS, Washington, D.C., and University of Arizona Press, Tuscon.

CAMERON, R.E. (1971)   In "Research in the Antarctic" (L.O. Quam and H.D. Porter, eds.), Publ. No. 93, 137. AAAS, Washington, D.C.

CAMERON, R.E. (1972a)   In "Antarctic Terrestrial Biology" (G.A. Llano, ed.), Antarctic Research Series 20, 195. American Geophysical Union, Washington, D.C.

CAMERON, R.E. (1972b) Phycologia 11, 133.

CAMERON, R.E. (1972c) In "Proceedings of the Colloquium
on Conservation Problems in Antarctica"
(B.C. Parker, ed.), 267. Allen Press, New York.

CAMERON, R.E. (1974) In "Polar Deserts and Modern Man"
(T.L. Smiley and J. H. Zumberge, eds.), 71.
University of Arizona Press, Tucson.

CAMERON, R.E., and BENOIT, R.E. (1970) Ecology 51, 802.

CAMERON, R.E., and FORD, A.B. (1974) Antarctic Journal of
the U.S. 9 , 116.

CAMERON, R.E., and MORELLI, F.A. (1974) Antarctic Journal
of the U.S. 9 , 113.

CAMERON, R.E., BLANK, G.B., and GENSEL, D.R. (1966) Jet
Propulsion Laboratory Tech. Rpt. 32-977, October
15, 1966, 153.

CAMERON, R.E., KING, J., and DAVID, C.N. (1970) In "Antarctic
Ecology" (M.W. Holdgate, ed.), 2, 702.
Academic Press, New York.

CAMERON, R.E., CONROW, H.P., GENSEL, D.R., LACY, G.H., and
MORELLI, F.A. (1971a) Antarctic Journal of the
U.S. 6, 211.

CAMERON, R.E., LACY, G.H., MORELLI, F.A., and MARSH, J.B.
(1971b) Antarctic Journal of the U.S. 6, 105.

CAMERON, R.E., MORELLI, F.A., and JOHNSON, R.M. (1972a)
Antarctic Journal of the U.S. 7, 187.

CAMERON, R.E., MORELLI, F.A., and RANDALL, L.P. (1972b)
Antarctic Journal of the U.S. 7, 254.

CAMERON, R.E., MORELLI, R., DONLAN, R., GUILFOYLE, J.,
MARKLEY, B., and SMITH, R. (1974a) Antarctic
Journal of the U.S. 9 , 141.

CAMERON, R.E., MORELLI, F.A., and HONOUR, R.C., (1974b) Dry Valley Drilling Project Bul. 4, 16.

HOROWITZ, N.H., CAMERON, R.E., and HUBBARD, J.S. (1972) Science 176, 242.

HUDEC, P.P., and SONNENFELD, P. (1974) Science 185, 440.

JAFFE, L.D., CAMERON, R.E., CHOATE, R., FENALE, F.P., HAINES, E., HOBBY, G.L., HUBBARD, J.S., MACKIN, R.J., SAUNDERS, R.S., and WHITEHEAD, A.B. (1974) Jet Propulsion Laboratory Document 760-101, May 28, 1974, 55.

MARTIN, S., Jr. (1972) NASA Langley Research Center M75-125-2, Rev. A. Apr. 14, 1972, 300.

MICHAUX, C.M. (1967) National Aeronautics and Space Administration, SP-3030 167.

MILTON, D.J. (1973) J. Geophys. Res. 78, 4037.

MILTON, D.J. (1974) Science 183, 654.

RUBIN, M.J. (1965) In "Biogeography and Ecology in Antarctica" (J. Van Miegham, P. Van Oye, and J. Schell, eds.), 72, Junk, The Hague.

RUBIN, M.J. (1970) Bul. of the Atomic Scientists 26, 48.

RUSIN, N.P. (1964) Meterological and Radiation Regime of Antarctica (In Russian). Gidrometeorologicheskoe Izdatelstvo, Leningrad, USSR, 1961. (English Translation Israel Program for Scientific Translation, Jerusalem.)

SAGAN, C. (1971) Icarus 15, 511.

SAGAN, C. (1972) Icarus 16, 1.

SAGAN, C. (1973) Science 181, 1045.

VISHNIAC, W.V. (1973). Unpublished Manuscript, 52.

VISHNIAC, W.V., and MAINZER, S.E., (1973) Life Sci. Space Res. 11, 25.

WILSON, A.T. (1970) <u>In</u> "Antarctic Ecology" (M.W. Holdgate,
        ed.), Vol. I, 21. Academic Press, New York.

YOUNG, R.S. (1973) <u>Space Life Sci.</u> <u>4</u>, 505.

YOUNG, R.S. (1974) Unpublished Manuscript, 3.

INFORMATION TRANSFER

INFORMATION TRANSFER IN THERMOPHILIC BACTERIA
J. Stenesh

COMPONENTS OF INFORMATION TRANSFER

INTRODUCTION

The phenomenon of thermophily, or growth at high temperatures, has intrigued scientists for many years. In the case of bacteria, thermophiles are usually considered to be organisms capable of surviving and growing at temperatures of about 55-80° C, as compared to mesophiles, which grow at temperatures of about 20-45° C (Bansum and Matney, 1965). Several theories have been advanced to explain the phenomenon of thermophily. According to Allen (1953), thermophily is a special metabolic state, character-ized by high rates of synthesis and degradation. Supportive of this theory are the experiments of Bubela and Holdsworth (1966), who found a more rapid turnover of proteins and nucleic acids at 40° C in Bacillus stearothermophilus, as compared to that in Escherichia coli. Brock (1967), on the other hand, concluded that thermophiles do not grow as rapidly at the higher temperatures as one would predict on the basis of the Arrhenius equation. A second theory (Bělehrádeck, 1931), considers thermophily to be due to the protective action of lipids, and attempts to correlate heat stability of the organism with the melting point of the cell lipid material. A number of workers have, indeed, found differences in the lipid composition of mesophiles and thermophiles (Kaneda, 1967; Daron, 1970; Shen *et al.*, 1970). A third theory ascribes thermophily to physical-chemical differences in the structure and function of important macromolecules from thermophiles, as compared to those from mesophiles (Koffler *et al.*, 1957; Friedman, 1968; Campbell and Pace, 1968).

To date, the latter theory has received the most support. The early work dealt mainly with studies of proteins, especially purified enzymes (Friedman, 1968; Campbell, 1955) and bacterial flagella (Koffler *et al.*, 1957), in which cases the macromolecules of the thermophiles were shown to possess an unusual stability toward denaturation by heat. The literature in this area has been reviewed by Gaughran (1947), Allen (1953), and Koffler (1957). About 10 years ago, thermophilic systems began to be studied in connection with their properties in information transfer; work on DNA, RNA, ribosomes, and other components of information transfer has increased rapidly, and advances resulting from this increase have been reviewed by Friedman (1968), and by Campbell and Pace (1968). At the same time, work has continued on the properties of purified proteins, and these studies have been reviewed by Singleton and Amelunxen (1973).

Many investigators have compared properties of macromolecules from thermophiles, generally B. stearothermophilus, with properties of corresponding macromolecules from the mesophile E. coli. Because these organisms belong to different genera, the possibility that some of the observed differences are due to intergeneric differences between the organisms cannot be ruled out. In order to avoid this complication, we have, in our laboratory, consistently studied both mesophilic and thermophilic organisms from the same genus, namely Bacillus.

DNA

DNA has been isolated from a number of thermophilic organisms (Campbell and Pace, 1968; Pace and Campbell, 1967; Saunders and Campbell, 1966b; Marmur, 1960; Stenesh *et al.*, 1968), including phages (Saunders and Campbell, 1965; Carnevali and Donelli, 1968; Egbert, 1969), lactobacilli (Dellaglio *et al.*, 1973), and an extreme thermophile, Thermus thermophilus (Oshima and Imahori, 1974). The guanine plus cytosine (G + C) content of these organisms varied from about 40 to 60 mole percent. The mesophilic strains examined in our laboratory had an average G+C content of 44.9 mole percent, whereas that of the thermophiles was 53.2 mole percent. The G+C content could be correlated with the maximum growth temperature of the organism. As expected, the DNA containing a greater G+C content was more stable to denaturation by heat (higher melting out temperature, or T_m), because of the more extensive hydrogen bonding that can take place between the individual strands.

RNA

Ribosomal RNA (rRNA) of thermophiles has likewise been shown to contain a higher G+C content than that of mesophiles (Pace and Campbell, 1967; Mangiantini *et al.*, 1965; Stenesh and Holazo, 1967). Pace and Campbell (1967) studied 19 organisms and found that generally the G+C content increased and the adenine plus uracil (A+U) content decreased as a function of increasing maximum growth temperature of the organism. The rRNA of an extreme thermophile, T. aquaticus, (Zeikus *et al.*, 1970), had a G+C content of 63.7 mole percent compared to 56.5 mole percent for the rRNA of E. coli, but only the 23S RNA and not the 16S RNA had a higher T_m than that of the E. coli RNA. Friedman *et al.* (1967) showed that B. stearothermophilus rRNA had a T_m of 60° C compared to one of 50° C for the RNA from E. coli. In our laboratory, the mesophilic strains of Bacillus had an average G+C content of 55.1 mole percent and an average T_m of 64.6° C compared to the corresponding values of 59.8 and 69.7 for the thermophiles (Stenesh and Holazo, 1967). Because the nucleotide maps of these RNAs were essentially identical in all cases, and because the thermal denaturation profiles showed two regions of melting out, the data indicated that the rRNAs from mesophiles and thermophiles had similar nucleotide sequences, but differed in the relative frequencies of these sequences.

Other types of RNA have also been isolated from thermophilic organisms. Saunders and Campbell (1966a) isolated messenger RNA (mRNA) from B. stearothermophilus, the G+C content of which correlated well with the base composition of the DNA. Stenesh *et al.* (1968) showed that total RNA from thermophilic strains had an average G+C content of 61.4 mole percent, compared to 56.9 mole percent for mesophilic strains. These values were higher than those for the rRNA, indicating that transfer RNA (tRNA) would be expected to be more stable than the rRNA. Recent studies

have borne this out. The tRNA of T. aquaticus (Zeikus et al.,
1970) had a T_m of 86° C, that of Flavobacterium thermophilum
(Oshima and Imahori, 1971) had a T_m of 87.5° C, compared to
a T_m of 80° C for the tRNA of E. coli. Leucine-tRNA of B.
stearothermophilus has been shown to consist of several
isoaccepting tRNAs and to have a higher T_m (64.5°C) than that
of the unfractionated tRNA (Lurquin et al., 1969). Since the
T_m was almost independent of wavelength (measured at 258 and
280 nm) and since the thermal denaturation profile was
monophasic, the authors concluded that the tRNA melted out
in an entirely cooperative fashion.

An interesting finding has been made by Agris et al.,
(1973), which showed that when B. stearothermophilus was
grown at both 50° and 70° C, 1.4 times as many methyl groups
were incorporated into the tRNA at the higher growth
temperature; this was due primarily to a three-fold increase
in the 2'-O-methylribose moieties of the tRNA. The types
and quantities of base-methylated nucleotides were almost
identical, whether the organism was grown at the lower or at
the higher temperature. The thermal denaturation profiles of
the two types of RNA were essentially identical, so that the
secondary structure of the tRNA does not appear to be
different. On the other hand, preliminary experiments by the
authors indicated that the increased methylation protects the
tRNA against nuclease attack, since the tRNA from the 70° C
cultures was digested by ribonuclease T_1 at about one-half
the rate as the tRNA from the 50° C cultures. In this
context, it is interesting to note the comparative study of
5S RNA, which seems to be devoid of methylated bases, from
B. stearothermophilus and E. coli by Gray and Saunders (1973).
The 5S RNAs did not differ in their chain length,
electrophoretic behavior, sedimentation, and chromatographic
properties, but the 5S RNA from the Bacillus species appeared
to be in a form that was less compact, and had greater
heterogeneity in types of base-paired regions. Valine and
isoleucine tRNA were isolated from B. stearothermophilus
(Johnson and Soell, 1971) and charged by reaction with the
aminoacyl-tRNA synthetase from the same organism; acceptor
activity of the tRNA was lost as the temperature was
increased, and this was attributed to alterations in the
secondary and tertiary structures of the tRNA, rather than
to the denaturation of the enzyme. This is in accord with
the finding of other workers, using B. stearothermophilus
(Arca et al., 1965), B. coagulans (Fresco et al., 1966), or

T. aquaticus (Zeikus and Brock, 1971), that a certain degree
of secondary structure in the tRNA is required for
aminoacylation by the enzyme.

RIBOSOMES

The thermal stability of thermophilic ribosomes has been
determined by a number of investigators (Pace and Campbell,
1967; Mangiantini et al., 1965; Zeikus et al., 1970;
Friedman et al., 1967). Pace and Campbell (1967) showed
that the T_m of ribosomes from 19 organisms correlated
positively with the maximum growth temperature of the
organisms. The melting-out of the ribosomes occurred at
higher temperatures than that of the corresponding ribosomal
RNA, so that the structural arrangement of the rRNA and the
ribosomal protein (r-protein) must be a contributing
stabilizing factor. B. stearothermophilus ribosomes have
been reported to have the same surface structure as mesophilic
ribosomes (Bassel and Campbell, 1969), but to be more stable
to degradation by associated ribosomal ribonuclease
(Mangiantini et al., 1965; Stenesh and Yang, 1967). Friedman
et al. (1967), and Algranati and Lengyel (1966) showed that
ribosomes from B. stearothermophilus, upon heating, retained
more of their capacity to support poly-U-dependent amino
acid incorporation than was the case for the ribosomes from
E. coli. The content of stabilizing polyamines in ribosomes
has been investigated by Stevens and Morrison (1968). Higher
thermal stability of ribosomes has also been reported for a
thermophilic species of Penicillium (Miller and Shepherd,
1973) and for thermophilic species of Clostridium (Irwin et
al. 1973), compared to mesophilic species from the same
genera.

Friedman (1971) has shown that both the 30S and the 50S
subunits of B. stearothermophilus ribosomes were more heat
stable than the corresponding components of E. coli.
However, reassociated 70S ribosomes did not differ in their
stability from reassociated E. coli ribosomes. Hybrid
ribosomes, consisting of subunits derived from B.
stearothermophilus and E. coli (Chang et al., 1966), as well
as those derived from rRNA and r-proteins of these organisms
(Nomura et al., 1968), have been shown to be active in
protein synthesis. The hybrid composed of 30S subunits from
E. coli and 50S subunits from B. stearothermophilus was more
heat stable than the reciprocal hybrid (Altenburg and
Saunders, 1971).

89

Because thermophiles contain such unusually heat stable ribosomes and other heat stable macromolecules, they are becoming increasingly recognized as a useful source of various components. Thus, the following have been isolated from B. stearothermophilus: elongation factor T_u (Skoultchi et al., 1968), a subparticle of the 30S ribosome (Chow et al., 1972), a 5S RNA-ribosomal protein complex (Horne and Erdmann, 1972), a 50S ribosomal particle reconstituted from protein-free RNA (Fahnestock et al., 1973), a heterologous complex of an E. coli r-protein and a 16S or a 23S rRNA of B. stearothermophilus (Daya-Grosjean et al., 1973), a 50S B. stearothermophilus ribosome reconstituted with either prokaryotic or eukaryotic 5S RNA (Wrede and Erdmann, 1973), initiation factors (Kay and Grunberg-Manago, 1972), and a ribosome association factor (Garcia-Patrone et al., 1971).

RIBOSOMAL PROTEIN

Refinements in the analysis of r-proteins, in general, have also led to a closer examination of r-proteins from thermophiles. Geisser et al. (1973) reported that B. stearothermophilus ribosomes contained a larger number of proteins than ribosomes from either mesophilic Bacillus species or E. coli (Horne and Erdmann, 1972; Fahnestock et al., 1973; Daya-Grosjean et al., 1973; Wrede and Erdmann, 1973; Kay and Grunberg-Manago, 1972). These authors also reported great heterogeneity among r-proteins from Bacillus species, both with respect to their electrophoretic properties and to their immunological cross-reactions with r-proteins from E. coli. On the other hand, the 30S r-proteins from B. stearothermophilus showed a high degree of correlation with those from E. coli, with respect to their electrophoretic mobility, immunological cross-reaction, and molecular weight (Isono et al., 1973), as well as with respect to sequence homologies (Higo and Loertscher, 1974; Yaguchi et al., 1973). Higo et al. (1973) showed that most of the E. coli r-proteins can be replaced by the corresponding, but chemically different, B. stearothermophilus proteins in the 30S ribosome reconstitution system, using poly-U-dependent polyphenylalanine synthesis as a criterion of ribosome functionality. These data suggest that ribosomes from taxonomically unrelated prokaryotic organisms may have the same fundamental structural organization.

ENZYMES

Several thermophilic enzymes involved in information transfer have been investigated. Leucyl- and isoleucyl-tRNA synthetases from B. stearothermophilus are heat-stable, and can aminoacylate tRNA *in vitro* at temperatures above 70° C (Vanhumbeeck and Lurquin, 1968); however, the aminoacyladenylate-enzyme complex does not differ greatly in its stability from that of the E. coli enzyme, (Charlier *et al.*, 1969). Arginyl-tRNA synthetase from B. stearothermophilus (Parfait and Grosjean, 1972) appears to catalyze amino acid activation in a concerted type mechanism in which the amino acid is bound to the enzyme, after ATP and tRNA have become bound to it, and in which all three substrates then react simultaneously. The presence of all three substrates on the enzyme leads to a synergistic protection of the enzyme against heat inactivation (Parfait, 1973). A DNA-dependent RNA polymerase has been isolated from both B. stearothermophilus (Remold-O'Donnell and Zillig, 1969) and T. aquaticus (Air and Harris, 1974). The enzyme from B. stearothermophilus had an optimum temperature of 47-51° C, depending on the bacterial strain, and was considerably more heat stable than the corresponding enzyme from E. coli. The enzyme from T. aquaticus had similar sub-units, optimum pH, and metal ion requirements as the RNA polymerase from E. coli, but had an optimum temperature of 50-60° C, and lost only 20% of its activity when incubated for 30 min. at 70° C.

In our laboratory, we have isolated a DNA-dependent DNA polymerase from B. stearothermophilus and B. licheniformis (Stenesh and Roe, 1972a). The polymerases had the same basic requirements for the reaction, but the polymerase from the thermophile required a higher magnesium ion concentration for maximal activity, and had an optimum temperature of 65° C, compared to one of 45° C for the mesophilic enzyme. Both homologous and calf thymus DNA could serve as templates for the reaction. In each case, activated DNA (deoxyribonuclease-treated) was the best template, followed by native DNA and heat-denatured DNA in this order. A DNA polymerase with an optimum temperature of 70° C has been isolated from T. thermophilus (Sakaguchi and Yajima, 1974).

IN VITRO PROTEIN-SYNTHESIZING SYSTEMS

Cell-free amino acid incorporating systems have been developed for a number of thermophilic organisms (Algranati and Lengyel, 1966; Friedman and Weinstein, 1966; Stenesh and Schechter, 1969; Stenesh *et al.*, 1971). Algranati and Lengyel (1966) used poly-U (1.4 : 1) as synthetic mRNA in a B. stearothermophilus system that incorporated optimally between 55–60° C. Friedman and Weinstein (1966) used a B. stearothermophilus system, and showed that optimum incorporation was also between 55–60° C. The incorporation was magnesium-dependent, such that at low magnesium ion concentrations (0.01 M), and at poly-U concentrations below 50 μg per incubation mixture, the incorporation at 37° C was greater than at 65° C, while at high magnesium ion concentrations (0.018 M), and at poly-U concentrations above 50 μg per incubation mixture, the incorporation at 65° C was greater than at 37° C. We have extended these findings in a study of B. stearothermophilus and B. licheniformis (Stenesh and Schechter, 1969; Stenesh *et al.*, 1971). For the thermophile, but not for the mesophile, the relative incorporation at 37° C and 55° C was reversed at a poly-U concentration of 80 μg per incubation mixture. Higher magnesium ion concentrations and lower ammonium ion concentrations tended to increase the incorporation at 55° C over that at 37° C in the B. stearothermophilus system, whereas they had the opposite effect in the B. licheniformis system. The effect of potassium ions paralleled that of ammonium ions. The relative incorporation at 37° C and 55° C was not affected by purification of either the ribosomes or the S-100 fraction, and was not affected by the type of tRNA used. Low concentrations of puromycin led to a stimulation of amino acid incorporation in these systems, while high concentrations of puromycin led to an inhibition (Stenesh and Shen, 1969). An amino acid incorporating system from B. stearothermophilus has been used by Dunlap and Rottman (1972), to test for the capacity of B. subtilis rRNA to act as a template at an incubation temperature at which some of its secondary structure has been lost by heat denaturation.

FIDELITY OF INFORMATION TRANSFER

INTRODUCTION

The fidelity of information transfer is of prime importance for the capacity of an organism to function properly under normal conditions and, as such, must be high in *in vivo* systems. On the other hand, deviations from fidelity may serve as a regulatory device to permit an organism to function under adverse or unusual conditions. The possibility therefore exists that such deviations in fidelity play a role in the adjustment of mesophilic and thermophilic organisms to their respective environments. Fidelity, or the lack of it, can be involved at one or more of the stages in the process of information transfer, that is, at the level of replication, transcription, and translation.

REPLICATION

That errors in replication can occur has been shown for both bacterial (Hall and Lehman, 1968; McCarter *et al.*, 1969) and mammalian (Springgate and Loeb, 1973; Chang, 1973) DNA-dependent DNA polymerases. In our laboratory, we have studied the fidelity of the replication of DNA from B. licheniformis and B. stearothermophilus, by either the homologous or the heterologous enzyme over a range of temperatures (37°-72° C), by means of nearest-neighbor frequency analysis of the reaction product (Stenesh and Roe, 1972b). It was shown that differences in the nearest-neighbor frequency analysis occurred as a function of temperature either when one DNA was replicated by the two polymerases or when the two DNAs were replicated by one polymerase. Hence, it was concluded that both the DNA template and the DNA-dependent DNA polymerase contribute toward producing the observed changes in the nearest-neighbor frequency analysis. The observed changes were somewhat more pronounced for B. stearothermophilus than for B. licheniformis, and were more pronounced for those dinucleotides where both bases are held together by two hydrogen bonds (A, T) than for those where the bases are held together by three hydrogen bonds (G, C). Although the data indicated the occurrence of errors, an alternative hypothesis, that different sections of the DNA are preferentially replicated at different temperatures, was entertained, but appeared to be ruled out on the basis of various experimental findings. We have since shown conclusively that errors in replication occur in these systems, by

determining the incorporation of labeled deoxyribonucleoside
triphosphates using either poly dAT or poly dG·poly dC
as a template (Stenesh and McGowan, in preparation).

TRANSCRIPTION

The fidelity of transcription has not been investigated
in thermophilic bacterial systems, but has been studied in
several other systems. Errors were shown to occur in the
transcription of DNA by DNA-dependent RNA polymerase of E.
coli (Strinste et al., 1973) and Micrococcus luteus
(Goddard et al., 1969); in these studies, increasing amounts
of noncomplementary bases were incorporated into the RNA
product as a function of the x-ray dose to which the enzyme
had been exposed. In other studies, the integrity of
Xenopus DNA was shown to affect the fidelity of its
transcription by DNA-dependent RNA polymerase (Hecht and
Birnstiel, 1972), and estrogen was shown to affect the
nearest-neighbor frequency distribution of the RNA
synthesized in rat uterus (Nicolette and Babler, 1972).
A significant lack of fidelity has also been reported for
the reverse transcriptase, which transcribes the RNA of
avian myeloblastosis virus (Springgate et al., 1973). It
is likely, therefore, that in thermophilic systems errors in
transcription may also occur as a result of alterations in
the properties of the DNA template and/or the transcribing
DNA-dependent RNA polymerase.

TRANSLATION

Lack of fidelity in translation can arise from a number
of sources. One such source is amino acid activation where
it is well known that mischarging can take place. Arca et
al. (1965) reported mischarging of isoleucyl-tRNA with
valine by isoleucyl-tRNA synthetase from B. stearothermophilus
at temperatures of 70°-75° C, but not at temperatures of
50°-60° C; the isoleucyl-tRNA was likewise mischarged with
serine and threonine at 75° C but not at 50° C. Okamoto
(1967) isolated a factor from B. stearothermophilus that
inhibited the binding of amino acids to transfer RNA.

Another source of errors in translation is described by
ambiguity, that is, the incorporation of one amino acid in
response to a codon for a different amino acid. Friedman
and Weinstein (1964) have shown that ambiguity in a B.
stearothermophilus system is affected considerably by the
conditions used during amino acid incorporation, such as

temperature, ionic composition, and nature of the solvent. In most cases, the ambiguity involved those amino acids in which the codons contained at least two bases that were also present in the copolymer that was used as a template (Friedman and Weinstein, 1965); the best-known ambiguity is that of the incorporation of leucine, instead of phenylalanine, in the presence of a poly-U template. An exception to the generalization was the incorporation of proline in response to a poly-UG template. The ambiguity in these studies was enhanced by low temperatures and high magnesium ion concentrations. Experiments in our laboratory (Stenesh *et al.*, in preparation) on B. licheniformis and B. stearothermophilus, confirm these findings; the highest leucine/phenylalanine ambiguity was obtained for the B. stearothermophilus system, and the ambiguity was more pronounced at 37° C than at 55° C for both organisms. Experiments with mixed fractions from B. licheniformis and B. stearothermophilus implicated the ribosomes as the major cause of the ambiguity. The ribosome is also considered to be the cause of the ambiguity in E. coli (Weinstein *et al.*, 1966; Friedman *et al.*, 1968) and in yeast (Schlanger and Friedman, 1973), and it has been suggested that the ribosome controls the accuracy of translation by having a tRNA-screening site that interacts with a portion of the tRNA molecule outside the anticodon region (Schlanger and Friedman, 1973; Gorini, 1971). It is conceivable that part or all of this tRNA screening site may consist of adsorbed substances, as washing of the ribosomes and the mode of ribosome preparation have been shown to affect the ambiguity in B. stearothermophilus (Chomczynski *et al.*, 1969; Perzynski *et al.*, 1969).

   The degeneracy of the genetic code is an additional area of variability in information transfer that should be mentioned, even though it does not constitute a lack of fidelity. The occurrence of several synonym codons for an amino acid (degeneracy) raises the possibility that thermophiles and mesophiles may differ in their relative utilization of synonym codons, particularly in view of the known differences in base composition of their respective DNAs. We have investigated this possibility by using cell-free amino acid incorporating systems from B. licheniformis and B. stearothermophilus in the presence of various polyribonucleotide copolymers (Stenesh *et al.*, in preparation). On the assumption that differences in the incorporation of a given amino acid in response to various polymeric templates can be attributed entirely to differences in utilization of synonym codons, the data suggest that (*a*) the mesophilic and the thermophilic

systems utilize the same major codons, (b) the thermophilic
system utilizes the same codons at both 37° C and 55° C, and
(c) the thermophilic system is more limited in its capacity
to use synonym codons. We have extended these experiments
by measuring the binding of charged aminoacyl tRNAs to
ribosomes in the presence of trinucleotide codons (Stenesh
et al., in preparation), using the ribosome binding assay.
The thermophilic system showed a pronounced preference for
certain synonym codons as a function of incubation temperature.

REFERENCES

AGRIS, P.F., KOH, H., and SOELL, D. (1973) Arch. Biochem.
Biophys. 154, 277.

AIR, G.M., and HARRIS, J.I. (1974) FEBS Letters 38, 277.

ALGRANATI, I.D., and LENGYEL, P. (1966) J. Biol. Chem. 241,
1778.

ALLEN, M.B. (1953) Bacteriol. Rev. 17, 125.

ALTENBURG, L.C., and SAUNDERS, G.F. (1971) J. Mol. Biol. 55,
487.

ARCA, M., FRONTALI, L., and TECCE, G. (1965) Biochim. Biophys.
Acta 108, 326.

BANSUM, H.T., and MATNEY, T.S. (1965) J. Bacteriol. 90, 50.

BASSEL, A., and CAMPBELL, L.L. (1969) J. Bacteriol. 98, 811.

BĚLEHRÁDECK, J. (1931) Protoplasma 12, 406.

BROCK, T.D. (1967) Science 158, 1012.

BUBELA, B., and HOLDSWORTH, E.S. (1966) Biochim. Biophys.
Acta 123, 364.

CAMPBELL, L.L. (1955) Arch. Biochem. Biophys. 54, 154.

CAMPBELL, L.L., and PACE, B. (1968) J. Appl. Bacteriol. 31,
24.

CARNEVALI, F., and DONELLI, G. (1968) Arch. Biochem. Biophys.
125, 376.

CHANG, F.N., SIH, C.J., and WEISBLUM, B. (1966) Proc. Natl.
Acad. Sci. U.S. 55, 431.

CHANG, L.M.S. (1973) J. Biol. Chem. 248, 6983.

CHARLIER, J., GROSJEAN, H., LURQUIN, P., VANHUMBEECK, J.,
    and WERENNE, J. (1969) FEBS Letters 4, 239.

CHOMCZYNSKI, P., PERZYNSKI, S., and SZAFRANSKI, P. (1969)
    Acta Biochim. Polonica 16, 379.

CHOW, C.T., VISENTIN, L.P. MATHESON, A.T., and YAGUCHI, M.
    (1972) Biochim. Biophys. Acta 287, 270.

DARON, H.H. (1970) J. Bacteriol. 101, 145.

DAYA-GROSJEAN, L., GEISSER, M., STOEFFLER, G., and GARRET,
    R.A. (1973) FEBS Letters 37, 17.

DELLAGLIO, F., BOTTAZZI, V., and TROVATELLI, L.D. (1973)
    J. Gen. Microbiol. 74, 289.

DUNLAP, B.E., and ROTTMAN, F. (1972) Biochim. Biophys. Acta
    281, 371.

EGBERT, L.N. (1969) J. Virology 3, 528.

FAHNESTOCK, S., ERDMANN, V., and NOMURA, M. (1973)
    Biochemistry 12, 220.

FRESCO, J., ADAMS, A., ASCIONE, R., HENLEY, D., and
    LINDAHL, T. (1966) Cold Spring Harbor Symp. Quant.
    Biol. 31, 527.

FRIEDMAN, S.M. (1968) Bacteriol. Rev. 32, 27.

FRIEDMAN, S.M. (1971) J. Bacteriol. 108, 589.

FRIEDMAN, S.M., and WEINSTEIN, I.B. (1964) Proc. Natl. Acad.
    Sci. U.S. 52, 988.

FRIEDMAN, S.M., and WEINSTEIN, I.B. (1965) Biochem. Biophys.
    Res. Commun. 21, 339.

FRIEDMAN, S.M., and WEINSTEIN, I.B. (1966) Biochim. Biophys.
    Acta 114, 593.

FRIEDMAN, S.M., AXEL, R., and WEINSTEIN, I.B. (1967) J.
    Bacteriol. 93, 1521.

FRIEDMAN, S.M., BEREZNEY, R., and WEINSTEIN, I.B. (1968)

J. Biol. Chem. <u>243</u>, 5044.

GARCIA-PATRONE, M., GONZALES, N.S., and ALGRANATI, I.D.
(1971) Proc. Natl. Acad. Sci. U.S. <u>68</u>, 2822.

GAUGHRAN, E.R.L. (1947) Bacteriol. Rev. <u>11</u>, 189.

GEISSER, M., TISHENDORF, G.W., and STOEFFLER, G. (1973)
Molec. Gen. Genet. <u>127</u>, 129.

GODDARD, J.P., WEISS, J.J., and WHEELER, C.M. (1969) Nature
<u>222</u>, 670.

GORINI, L. (1971) Nature New Biology <u>234</u>, 261.

GRAY, P.N., and SAUNDERS, G.F. (1973) Arch. Biochem. Biophys.
<u>156</u>, 104.

HALL, Z.W., and LEHMAN, I.R. (1968) J. Mol. Biol. <u>36</u>, 321.

HECHT, R.M., and BIRNSTIEL, M.L. (1972) Eur. J. Biochem. <u>29</u>,
489.

HIGO, K.I., and LOERTSCHER, K. (1974) J. Bacteriol. <u>118</u>, 180.

HIGO, K., HELD, W., KAHAN, L., and NOMURA, M. (1973) Proc.
Natl. Acad. Sci. U.S. <u>70</u>, 944.

HORNE, J.R., and ERDMANN, V.A. (1972) Molec. Gen. Genet. <u>119</u>,
337.

IRWIN, C.C., AKAGI, J.M., and HIMES, R.H. (1973) J. Bacteriol.
<u>113</u>, 252.

ISONO, K., ISONO, S., STOEFFLER, G., VISENTIN, L.P., YAGUCHI,
M., and MATHESON, A.T. (1973) Molec. Gen. Genet.
<u>127</u>, 191.

JOHNSON, L., and SOELL, D. (1971) Biopolymers <u>10</u>, 2209.

KANEDA, T. (1967) J. Bacteriol. <u>93</u>, 894.

KAY, A.C., and GRUNBERG-MANAGO, M. (1972) Biochimie <u>54</u>, 1281.

KOFFLER, H. (1957) Bacteriol. Rev. <u>21</u>, 227.

KOFFLER, H., MALLETT, G.E., and ADYE, J. (1957) Proc. Natl.
Acad. Sci. U.S. <u>43</u>, 464.

LURQUIN, P., METZGER, P., and BUCHET-MAHIEU, J. (1969).
FEBS Letters 5, 265.

MANGIANTINI, M.T., TECCE, G., TOSCHI, G., and TRENTALANCE,
A. (1965) Biochim. Biophys. Acta 103, 252.

MARMUR, J. (1960) Biochim. Biophys. Acta 38, 342.

MCCARTER, J.A., KADOHAMA, N., and TSIAPALIS, C. (1969)
Can. J. Biochem. 47, 391.

MILLER, H.M., and SHEPHERD, M.G. (1973) Can. J. Microbiol.
19, 761.

NICOLETTE, J.A., and BABLER, M. (1972) Arch. Biochem. Biophys.
149, 183.

NOMURA, M., TRAUB, P., and BECHMANN, H. (1968) Nature 219,
793.

OKAMOTO, T. (1967) Biochim. Biophys. Acta 138, 198.

OSHIMA, T., and IMAHORI, K. (1971) J. Gen. Appl. Microbiol.
17, 513.

OSHIMA, T., and IMAHORI, K. (1974) J. Biochem. 75, 179.

PACE, B., and CAMPBELL, L.L. (1967) Proc. Natl. Acad. Sci.
U.S. 57, 1110.

PARFAIT, R. (1973) Eur. J. Biochem. 38, 572.

PARFAIT, R., and GROSJEAN, H. (1972) Eur. J. Biochem. 30, 242.

PERZYNSKI, S., CHOMCZYNSKI, P., and SZAFRANSKI, P. (1969)
Biochem. J. 114, 437.

REMOLD-O'DONNELL, E., and ZILLIG, W. (1969) Eur. J. Biochem.
7, 318.

SAKAGUCHI, K., and YAJIMA, Y. (1974) Federation Proc. 33,
1492.

SAUNDERS, G.F., and CAMPBELL, L.L. (1965) Biochemistry 4,
2836.

SAUNDERS, G.F., and CAMPBELL, L.L. (1966a) J. Bacteriol. 91,
332.

SAUNDERS, G.F., and CAMPBELL, L.L. (1966b) J. Bacteriol. 91, 340.

SCHLANGER, G., and FRIEDMAN, S.M. (1973) J. Bacteriol. 115, 129.

SHEN, P.Y., COLES, E., FOOTE, J.L., and STENESH, J. (1970). J. Bacteriol. 103, 479.

SINGLETON, JR. R., and AMELUNXEN, R.E. (1973) Bacteriol. Rev. 37, 320.

SKOULTCHI, A., ONO, Y., MOON, H.M., and LENGYEL, P. (1968) Proc. Natl. Acad. Sci. U.S. 60, 675.

SPRINGGATE, C.F., and LOEB, L.A. (1973) Proc. Natl. Acad. Sci. U.S. 70, 245.

SPRINGGATE, C.F., BATTULA, N., and LOEB, L.A. (1973) Biochem. Biophys. Res. Commun. 52, 401.

STENESH, J., and HOLAZO, A.A. (1967) Biochim. Biophys. Acta 138, 286.

STENESH, J., and MCGOWAN, G.R. (Manuscript in preparation).

STENESH, J., and ROE, B.A. (1972a) Biochim. Biophys. Acta 272, 156.

STENESH, J., and ROE, B.A. (1972b) Biochim. Biophys. Acta 272, 167.

STENESH, J., and SCHECHTER, N. (1969) J. Bacteriol. 98, 1258.

STENESH, J., and SHEN, P.Y. (1969) Biochem. Biophys. Res. Commun. 37, 873.

STENESH, J., and YANG, C. (1967) J. Bacteriol. 93, 930.

STENESH, J., ROE, B.A., and SNYDER, T.L. (1968) Biochim. Biophys. Acta 161, 442.

STENESH, J., SCHECHTER, N., SHEN, P.Y., and YANG, C. (1971) Biochim. Biophys. Acta 228, 259.

STENESH, J. SCHECHTER, N., and SHARMA, S.D. (Manuscript in preparation).

STEVENS, L., and MORRISON, M.R. (1968) Biochem. J. 108, 633.

STRINSTE, G.F., SMITH, D.A., and NEWTON-HAYES, F. (1973) Biochemistry 12, 603.

VANHUMBEECK, J., and LURQUIN, P. (1968) Biochem. Biophys. Res. Commun. 31, 908.

WEINSTEIN, I.B., FRIEDMAN, S.M., and OCHOA, JR., M. (1966) Cold Spring Harbor Symp. Quant. Biol. 31, 671.

WREDE, P., and ERDMANN, V.A. (1973) FEBS Letters 33, 315.

YAGUCHI, M., ROY, C., MATHESON, A.T., and VISENTIN, L.P. (1973) Can. J. Biochem. 51, 1251.

ZEIKUS, J.G., and BROCK, T.D. (1971) Biochim. Biophys. Acta 228, 736.

ZEIKUS, J.G., TAYLOR, M.W., and BROCK, T.D. (1970) Biochim. Biophys. Acta 204, 512.

TEMPERATURE EFFECTS ON TRANSFER RNA

Brian R. Reid

NUCLEAR MAGNETIC RESONANCE PROPERTIES OF NUCLEIC ACIDS

High resolution NMR is a spectroscopic tool that can yield valuable information about the conformational dynamics and structure of transfer RNA (tRNA) in aqueous solution. The protons involved in hydrogen bonds in Watson-Crick base pairs exchange with solvent water protons; Englander and Englander (1965) showed that the exchange half-life of protons in the hydrogen bonds of tRNA base pairs, in aqueous solution, was several seconds; whereas proton lifetimes of only about 5 msec in a given environment are required to generate a discrete NMR characteristic of that environment (Becker, 1969). NMR investigations of model nucleosides in nonaqueous solvents by Katz and Penman (1966) revealed that, although the exocyclic amino proton resonances were quite close to the resonance position of water, the ring nitrogen proton of guanine (G) and uracil (U) resonated at much lower field. This observation, combined with the slow exchange data on base-paired protons, led to the pioneer experiments of Kearns, Patel, and Shulman (1971); in these experiments the low field proton spectrum of tRNA was studied in H_2O solutions. Solubility problems limited these studies to 1-2 mM tRNA solutions (25-50 mg/ml) in a water solvent of about 110 M in protons, hence the enormous water resonance at about -5 ppm from 2,2-dimethyl-2-silapentane-5-sulfonate (DSS) extended from about -10 ppm to 0 ppm. Despite this

103

difficult technical problem, discrete resonances were observed
by signal averaging in the -10 to -15 ppm region which, by
virtue of their large chemical shifts, could only be derived
from hydrogen-bonded ring NH protons in base pairs (the ring
NH protons of single-stranded residues exchange very rapidly
with solvent protons, and effectively time-average in the
water peak). Hence, in the Watson-Crick base pairs shown in
Figure 1, only the ring NH-ring N proton, shown in boldface,
generates a resonance in the low field (-10 to -15 ppm) region.

Figure 2 shows the low field NMR spectrum of highly purified yeast tRNA^{Phe}, which we began studying in detail in 1971. The spectrum shows several resolved peaks of varying intensity in the -11 to -15 ppm region which, for reasons previously explained, can only be derived from the ring NH-- ring N hydrogen bonds of complementary base pairs. Since each Watson-Crick pair contains only one such bond, it is apparent that this region of the spectrum contains a single "reporter" resonance from each base pair which exists with a lifetime longer than 5 msec in the tRNA molecule in solution. It is obvious that the assignment of each individual base pair to a specific resonance would result in an enormously powerful tool with which to monitor the dynamics and structure of all of the helical regions in intact tRNA.

Before we discuss the methods used to assign resonances, it is, perhaps, advisable to consolidate the information obtained so far, in order to establish that the observed resonances really do obey the criteria expected for ring NH-- ring N hydrogen bonds of complementary base pairs. Since these are solvent exchangeable (albeit slowly) protons, their resonance should disappear after extensive H-D exchange in D_2O solvents. This was found to be the case (Wong et al., 1972). Furthermore, the total proton intensity in the -11 to -15 ppm region should equal the number of complementary base pairs. Attempts to calibrate the total intensity of the spectrum with respect to the methyl resonance of the external standard Metcyanomyoglobin (MetCNMb) led to a value of 21 ± 3 protons (Wong et al., 1972). This is quite close to the 20 Watson-Crick base pairs in the cloverleaf, secondary structure of this tRNA. It is obvious that the various peaks in the spectrum are not of equal intensity, and thus, several peaks must contain resonances from more than one base pair. Since each peak should have an integral intensity, attempts were made to refine the integration accuracy by internal calibration, with respect to a resolved resonance. The B,C peak at -13.7 ppm corresponds to 16% of the total intensity, and thus has an intensity of approximately 3.4 protons. On

---

Fig. 1. Hydrogen bonding in complementary base pairs. The ring NH--ring N bond which is responsible for the low field NMR resonance is shown in boldface.

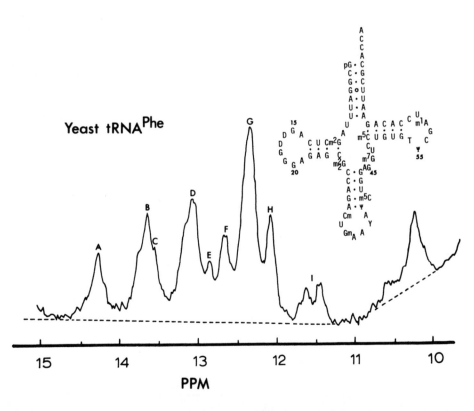

Fig. 2. *The low field NMR spectrum at 300 MHz of pure yeast tRNA^{Phe} at 35°C in 0.1 M NaCl, 0.01 M MgCl$_2$, 0.01 M sodium cacodylate pH 7.0.*

the assumption that this corresponds to 3.0 protons, the total
spectrum was found to contain slightly less than 19 base
pairs, i.e., one less than predicted from the cloverleaf
structure. A note of caution should be added here: if the
13.7 ppm resonance contains more than 3 protons, then the
total spectrum contains several extra base pairs (perhaps
from tertiary structure?) above that predicted from the cloverleaf secondary structure. However, this ambiguity is not
the central topic of this paper, and is not crucial to the
later interpretation of other tRNA spectra.

RESONANCE ASSIGNMENTS

     In order to determine the resonance positions of the
individual base pairs, we isolated the individual hairpin
helices of yeast tRNA^{Phe} using controlled endonuclease clea-
vage conditions, which we have previously published
(Schmidt *et al.*, 1970; Reid *et al.*, 1972). These four and
five base pair helices gave simple NMR spectra of the corresponding ring NH--ring N protons, as shown in Figure 3, and greatly
simplified assignment of the resonances (Lightfoot *et al.*,
1973). Using such an approach, we were able to make the
assignments shown in Figure 4. An interesting aspect of this
result was the observation that the resonance position of a
given guanine-cytosine (GC) or adenine-uracil (AU) pair was
determined by its nearest neighbors on either side. R. G.
Shulman suggested that this phenomenon was a ring current
shift effect by the neighboring bases in the helical stack,
and was able to determine the numerical value of the upfield
ring current shift, using a "best fit" approach to the
observed resonance positions (Shulman *et al.*, 1973). These
shift values, although accurate to only approximately 0.3
ppm, have been extremely useful in predicting and assigning
resonances in subsequent tRNA spectra.

TEMPERATURE EFFECTS

     An interesting observation in the spectrum of the five
base pair anticodon helix shown in Figure 3, was the fact
that the three peaks of ratio 2:2:1 revealed five base pairs
only at temperatures of 37°C or below. At 45°C, one of the
resonances disappeared reversibly, a phenomenon which, initially, seemed somewhat puzzling. However, the explanation
of this disappearance lies in our initial discussion of
exchange kinetics in NMR, where we mentioned that a resonance
will be observed only if the proton lifetime in the helix is
longer than about 5 msec. Thus, warming a sample to the

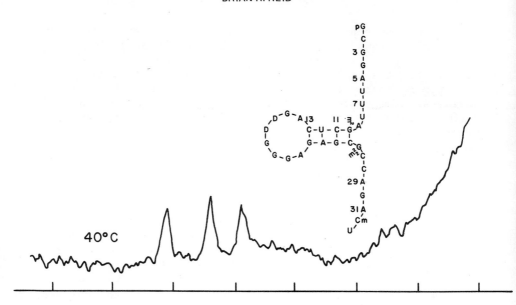

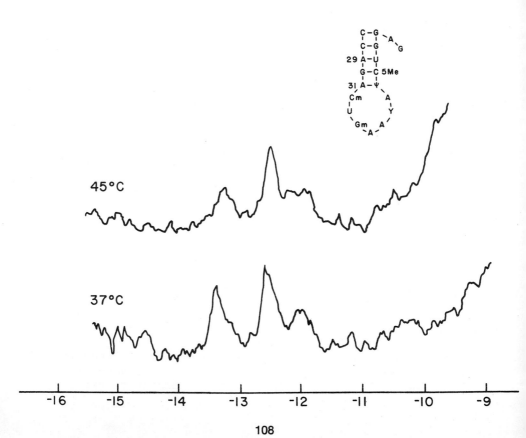

point at which a base pair helical lifetime becomes less
than this value, will cause the resonance to be "washed out"
into the water resonance, due to fast exchange of the non-
helical form with solvent protons. This occurs for one of
the base pairs (a terminal one) between 37°C and 45°C. More
detailed experiments with such helical hairpin molecules
have shown progressive sequential "fraying" of the helix from
the ends, as the temperature is raised.

The loss of a resonance is referred to as "melting"; but
NMR melting should not be confused with the more familiar
optical melting, as these are separate processes. At room
temperature, the helix opening rate is slow (about a few
seconds), whereas the closing rate is fast (usually faster
than 1 msec). At the optical melting point (T_m), the helix-
coil transition is at equilibrium with equally populated
helical and coil states; hence, the opening rate must equal
the closing rate. Thus the effect of heating the sample to
its optical T_m is to accelerate the opening rate until it
catches up with, and equals, the closing rate (usually in the
μsec-msec time range). Before this is attained, however, the
accelerating opening rate will pass through the 5 msec time
range, causing loss of the resonance. Hence, "NMR melting"
is usually observed at temperatures several degrees lower
than the optical T_m.

TEMPERATURE-DEPENDENT CHANGES IN THE DYNAMICS AND STRUCTURE
OF tRNA

Since discrete resonances from isolated helices can be
lost from the NMR spectrum when their helix-coil exchange
rate becomes sufficiently rapid, it was of interest to study
the NMR spectrum of intact tRNA species at various temperatures, to determine if the component helices in the tRNA
exhibited different dynamic properties. Studies on intact
yeast tRNA^{Phe} revealed a sudden cooperative melting of all
resonances over a small temperature range around 70°C, with
no partial melting at intermediate temperatures. However, a

---

*Fig. 3. The low field NMR spectrum of the DHU helix (top),
and the anticodon helix (bottom) of yeast tRNA^{Phe};
the solvent is the same as in Figure 2.*

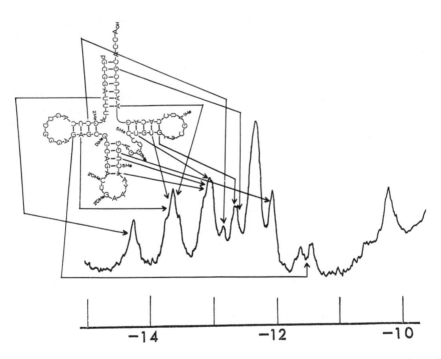

Fig. 4. *Assignment of resonances to individual base pairs in yeast tRNA*^{Phe}*. The split resonance at -11.5 ppm has been tentatively assigned to GC13 in the DHU stem, however a note of caution should be added, in that no resonance at -11.5 ppm was observed in the isolated DHU helix containing GC13. The remaining unassigned GC pairs generate the large peak at -12.3 ppm.*

temperature-dependent study of $\underline{E}$. $\underline{coli}$ tRNA^{Tyr} did reveal discrete losses in the NMR spectrum at intermediate temperatures. The spectrum of tyrosine tRNA in 0.1 M NaCl, 10 mM Mg^{2+} at 31°C is shown at the top of Figure 5; the spectrum contains

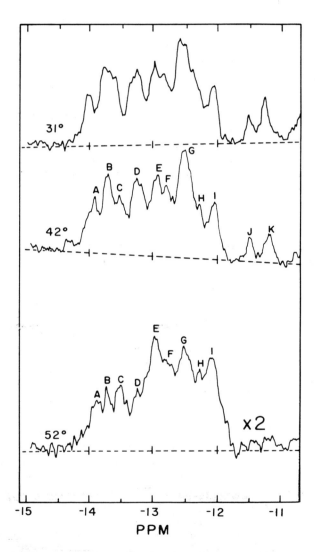

Fig. 5.   The 300 MHz low field NMR spectrum of $\underline{E}$. $\underline{coli}$ tRNA^{Tyr} in 0.1 M NaCl, 0.01 M MgCl$_2$, 0.01 M sodium cacodylate pH 7.0.

approximately 24 base pairs, and very little change is observed
upon raising the temperature to 42°C. However, a further
increase to 52°C produces a marked change in the spectrum and
a reduction of intensity to approximately one-half, indicating
12-13 base pairs with lifetimes greater than 5 msec at this
temperature. Previous studies by Yang *et al.* (1972) reported
that E. coli tRNA^{Tyr} dimerizes to a functionally inactive form
upon heating to 50°C in 0.5 M Na^{+} buffers; the material from
the 52°C spectrum in Figure 5 was subjected to analysis by
aminoacylation, 4thioU hyperchromicity at 340 nm, and molecular weight, and was indeed shown to be the dimer. Based on
their optical studies, Yang *et al.* proposed a structure for
the dimer in which the rT helix was unfolded (as was the DHU
helix), and used in intermolecular base pairing. We decided
to study the resonances remaining in the dimer at intermediate
temperatures, to see if the base pairs from which they were
derived were consistent with their proposed dimer structure.
The major spectral change upon dimerization, i.e., between
42° and 52°C, is the loss of intensity from peaks A (1 proton),
B (2 protons), C (1 proton), D (2 protons), G (2-3 protons),
with peaks E, F, H, and I remaining unchanged; peaks J and K
are lost at 52°C.

   This demonstration of partial melting prompted a study
in the absence of Mg^{2+} at 0.01 M Na^{+} (conditions under which
no dimerization occurs), to determine which helices were the
last to unfold. As shown in Figure 6, there is a loss of
intensity around -13.8 and -13.2 ppm at 42°C. The most dramatic intensity change occurs upon raising the temperature to
52°C or 62°C. At 62°C, the six peaks containing seven protons
can only be accounted for by the seven base pair acceptor
helix, for example, UA3 (predicted -13.6 ppm; observed
-13.5 ppm), GC2 (predicted -13.2 ppm; observed -13.0 ppm); GC5
(predicted -12.6 ppm; observed -12.6 ppm); GC6 (predicted
-12.6 ppm; observed -12.5 ppm); GC1 (predicted -11.9 ppm;
observed -12.0 ppm); and GC4 (predicted -11.9 ppm; observed
-12.0 ppm). GC7 is predicted at -13.3 ppm if only GC6 contributes an upfield shift, or at -12.7 ppm if GC58 also contributes an upfield shift from below. Since the only unassigned
resonance is at -12.8 ppm, the latter must be the case indicating that the rT helix is stacked below the acceptor helix
(the stacking of the acceptor helix and the rT helix upon
each other is a feature seen in the low resolution crystal
structure of yeast tRNA^{Phe}). The intensity loss between 52°C
and 62°C is close to that predicted for the melting of the rT
helix. Thus, at 62°C the rT helix base pairs are too short-
lived to generate resonances, but GC58 is still structurally

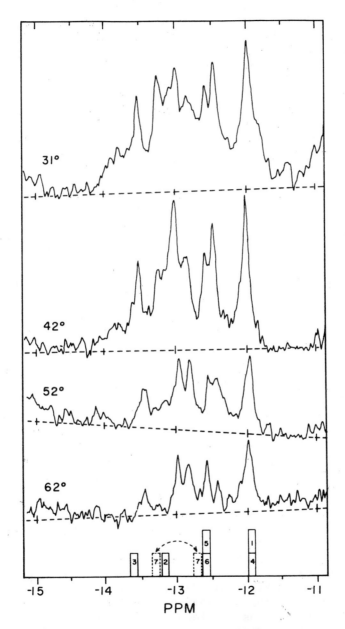

Fig. 6. *The low field 300 MHz NMR spectrum of E. coli* tRNA^*Tyr* *in 1 mM EDTA, 5 mM sodium cacodylate pH 7.0 at a total solvent Na⁺ concentration of 10 mM.*

present from the observed position of the GC7 resonance.

This melting pattern indicates that the acceptor helix is the most stable helix, and the rT helix is the second last helix to unfold. The fact that the rT helix appears more stable than the anticodon helix is corroborated by thermo-dynamic stability calculations (Tinoco et al., 1973), which indicate -6.6 kcal for the rT helix and -5.6 kcal for the anticodon helix. Since the acceptor helix and rT helix are stable even at low ionic strength up to 50°-60°C, they are likely to be even more stable at higher ionic strength in the 45°-50°C range where dimerization occurs. Thus, during dimerization it is very unlikely that the acceptor helix or the rT helix can unfold, and participate in bimolecular base pairing. This leaves the region between residues 8 and 57 as the region in which bimolecular base pairing must occur in the dimer. The matrix showing all possible base pairs between residues 8-57, is shown in Figure 7. Both halves of this symmetrical matrix are shown to reveal bimolecular, as well as unimolecular, helices. The only five base pair helix is that involving residues 28-32 with residues 44-40, i.e., the anticodon helix; for entropic reasons we feel this is more likely to pair in a unimolecular rather than a bimole-cular fashion, even though its base pairs are too short-lived to generate resonances at 52°C. The less stable three base paired minor stem and "DHU stem" may fold in an intra-molecular or intermolecular helix, however bimolecular pair-ing of the minor stem would eliminate the severe strain caused by a three nucleotide hairpin loop (Tinoco et al., 1973), thus favoring intermolecular bonding of these bases (46-48 with 54-52). If the DHU stem (10-12 and 26-24) reforms by intramolecular pairing, then there appear to be no extra pairs in the dimer, over and above those in the monomer, with which to drive the dimerization process. However, if the DHU stem refolds by intermolecular pairing then a new helix con-taining four GC pairs (17-20 with 20-17) can be formed in the dimer, which cannot form in the monomer. These four extra bimolecular GC pairs are sufficient to drive the reaction in the direction of dimerization, at temperatures sufficiently high to unfold the DHU stem. Thus, we envisage the bimole-cular base pairing in the dimer to involve the minor stem, the DHU stem and residues 17-20, i.e., a total of 16 base pairs. Although the NMR data give no indication concerning a unimolecular versus bimolecular anticodon helix, we prefer the former, since dimerization occurs at only 45°-50°C during the initial 20% of the UV hyperchromicity, suggesting very little unfolding of monomer helices. Hence, maintaining the

114

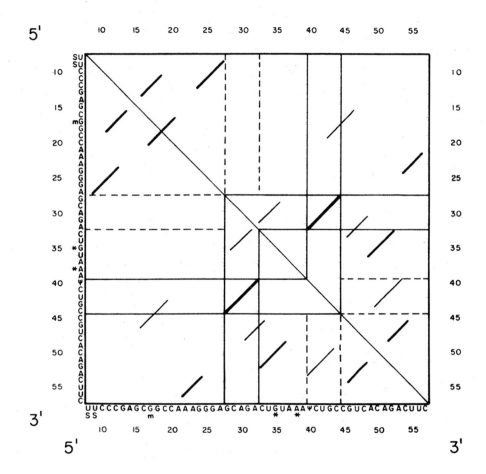

Fig. 7. *Matrix showing possible complementary base pair for-*
mation in the region from residue eight to residue
57 in E. coli tRNA^{Tyr}. The symmetry about the dia-
gonal line reflects the fact that helices can form
with the complementary sequence from the same mole-
cule, or from a second molecule. The five base pair
anticodon helix is shown as a thickened line. The
extended rectangles indicate that sequences used in
forming the anticodon helix cannot also participate
in other helices, either intramolecularly or inter-
molecularly. Only helices of three or more base
pairs are shown.

integrity of the acceptor stem, the rT stem and the anticodon
stem, and disrupting only 6 of the 23 secondary base pairs
(the minor stem and the DHU stem), is consistent with this
observation.   Furthermore, 16 bimolecular base pairs also
appears to be consistent with the 65°C temperature requirement
for remonomerization of the dimer at high dilution (Yang *et al.*,
1972).

The proposed dimer structure is shown in two-dimensional
representation in Figure 8.   Three-dimensional wire backbone
models can be built without serious steric restraints.

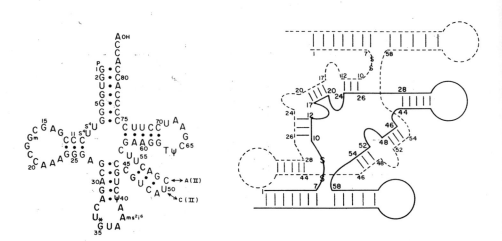

*Fig. 8.   The monomer cloverleaf of E. coli tRNA^{Tyr} and the
proposed intermolecular base pairing in the dimer.
One molecule is shown as a solid line, and the
second molecule in the dimer is shown as a dotted
line.*

ACKNOWLEDGMENTS

My thanks to Y. P. Wong for running the 300 MHz spectra
at Riverside, and to D. R. Kearns and B. F. Rordorf for critical discussion of the proposed dimer structure.   Grants from
the NSF (BMS73-01675) and USPHS (CA 11697) are gratefully
acknowledged.

REFERENCES

BECKER, E. D. (1969). "High Resolution NMR." Academic Press, New York.

ENGLANDER, S. W., and ENGLANDER, J. J. (1965). Proc. Nat. Acad. Sci. U.S.A. 53, 370.

KATZ, L., and PENMAN, S. (1966). J. Mol. Biol. 15, 220.

KEARNS, D. R., PATEL, D., and SHULMAN, R. G. (1971). Nature 229, 338.

LIGHTFOOT, D. R., WONG, K. L., KEARNS, D. R., REID, B. R., and SHULMAN, R. G. (1973). J. Mol. Biol. 78, 71.

REID, B. R., EINARSON, B. L., and SCHMIDT, J. (1972). Biochimie 54, 325.

SCHMIDT, J., BUCHARDT, B., and REID, B. R. (1970). J. Biol. Chem. 245, 5743.

SHULMAN, R. G., HILBERS, C. W., KEARNS, D. R., REID, B.R., and WONG, Y. P. (1973). J. Mol. Biol. 78, 57.

TINOCO, I., BORER, P. N., DENGLER, B., LEVINE, M. D., UHLENBECK, O. C., CROTHERS, D. M., and GRALLA, J. (1973). Nature New Biology 246, 40.

WONG, Y. P., KEARNS, D. R., REID, B. R., and SHULMAN, R. G. (1972). J. Mol. Biol. 72, 725.

YANG, S. K., SOLL, D. G., and CROTHERS, D. M. (1972). Biochemistry 11, 2311.

INFORMATION TRANSFER: SALT EFFECTS

S. T. Bayley

INTRODUCTION

Extremely halophilic bacteria are those which require 20-30% NaCl for optimal growth. So far, all bacteria that have been shown unequivocally to be extremely halophilic by this criterion belong to one of two genera, namely, Halobacterium or Halococcus (Larsen, 1967). This chapter is concerned entirely with halobacteria, the more extensively studied of the two genera.

Direct measurements on the internal concentrations of salt in both moderately and extremely halophilic bacteria by Christian and Waltho (1962) show that under optimal growth conditions, these bacteria contain concentrations of salt roughly comparable to those in the external media. For Halobacterium salinarium, grown on a medium containing 4 M NaCl and 0.03 M KCl, the internal concentration of chloride ions was similar to that in the medium, while the total concentration of monovalent cations was significantly greater, being 1.4 molal in Na^+ and 4.6 M in K^+. Clearly, therefore, the whole internal machinery of the halobacteria must have adapted to, or at least become tolerant of, unusually high concentrations of salt. In fact, as will be evident from results presented in this and other papers in this Symposium, halobacteria are obligate halophiles, having adapted irreversibly to an environment of concentrated salt. These bacteria require at least 12-15% NaCl for growth; at 1 *M* (about 3%) NaCl they rapidly lose viability, and at lower concentrations they lyse (Abram and Gibbons, 1960).

THE DNA OF EXTREMELY HALOPHILIC BACTERIA

Before discussing the detailed biochemistry of information transfer, we shall first consider what is known of the DNA in extremely halophilic bacteria.

Joshi *et al.*, (1963) showed by isopycnic centrifugation on CsCl density gradients that two species of halobacteria each contained two DNA components. In both cases, the main component had the higher density corresponding to a guanine-cytosine (GC) content of 67%, while the minor component had a density corresponding to a GC content of 59%. Later, Moore and McCarthy (1969a,b) confirmed these observations and found, further, that all of the extremely halophilic bacteria they examined, whether halobacteria or halococci, contained two DNA species of similar densities. The only difference between bacterial species was in the amount of the minor, satellite component they contained. This ranged from 11% to 36% of the total DNA.

The GC content of the main component is high, but not excessively so in comparison to some other bacteria (Singer and Ames, 1970). Singer and Ames (1970) suggested that high GC contents resulted from evolutionary pressure caused by exposure of bacteria to ultraviolet radiation. This, rather than adaptation to concentrated salt, may be the reason for a high GC content in the halobacteria, which normally grow under intense sunlight, e.g., in salt pans.

Studies by Moore and McCarthy (1969a) and in my own laboratory (Lou, 1970) indicate that the satellite DNA is a distinct component of extremely halophilic bacteria, and is not due to contaminating organisms or to artifacts of preparation. From the kinetics of renaturation of the DNA, Moore and McCarthy (1969b) concluded that the satellite cannot represent multiple copies of a small episomal element. These authors estimated from their renaturation curves that the genome size of Halobacterium spp. is 4.1×10^6 nucleotide pairs, compared to 4.5×10^6 for E. coli. In H. salinarium, the satellite represents 19% of the total DNA, corresponding to 8×10^5 nucleotide pairs. From the relative change in buoyant density in CsCl gradients caused by adding ethidium bromide, Lou (1970) found that the satellite DNA of H. salinarium behaved as closed circular molecules. In the electron microscope he observed that preparations of this material contained circles with a mean circumference of 37 μm. This length corresponds to about 110,000 nucleotide pairs, suggesting that each cell of H. salinarium may contain seven or eight of these molecules. The variation in amount of satellite between Halobacterium species could, therefore, be due to different numbers of such molecules per cell, although the renaturation results of Moore and McCarthy (1969b) suggest that each of these molecules in the cell would represent a unique nucleotide sequence. It is not clear what role this satellite DNA plays in extremely halophilic bacteria, or why all of these bacteria contain it.

TRANSCRIPTION

The most extensive study on a DNA-dependent RNA polymerase of an extreme halophile has been that of Louis and Fitt, working with H. cutirubrum. These authors (Louis and Fitt, 1971a,b; 1972a,b; Louis et al., 1971) describe an enzyme composed of two subunits, α and β, each of molecular weight 18,000 daltons, as estimated by gel filtration and density gradient centrifugation. The α-subunit carries the catalytic site for the synthesis of RNA, while the β-subunit controls the specificity of chain initiation (Louis and Fitt, 1972b). The complete enzyme, which consists of α and β subunits in a 1:1 complex requiring Mn^{2+} ions, is sensitive to rifampicin (Louis and Fitt, 1972b). The specificity of the complete enzyme is both salt-dependent and remarkably high. Thus DNA from calf thymus is transcribed only in dilute salt solution, whereas DNA from H. cutirubrum is transcribed only in concentrated salt. Furthermore, the complete enzyme transcribes the early genes preferentially and asymmetrically in native, double stranded DNA from T7 bacteriophage, as does E. coli RNA polymerase (Louis and Fitt, 1972a).

Independently, we have described a DNA-dependent RNA polymerase from H. cutirubrum that has significantly different properties (Chazan and Bayley, 1973). This enzyme was prepared by treating a DNA-protein-membrane complex from the cell homogenate with DNase, and then separating the polymerase from the nuclease by gel chromatography. From gel filtration and from sedimentation through density gradients, the molecular weight of this enzyme was estimated to be 300,000-400,000 daltons. As part of the original DNA-protein-membrane complex, the RNA polymerase was active at all concentrations of KCl or NH_4Cl up to saturation. After separation from the complex, the enzyme required added DNA template, but under these circumstances it was active only at salt concentrations below 0.4 *M*. This enzyme showed no template specificity and was not inhibited by rifampicin.

From the considerable differences in properties of their preparations, the two laboratories appear to be working with different enzymes (Chazan and Bayley, 1973). If this is the case, the question then arises of the interrelationship in function between two RNA polymerases in the same prokaryotic cell. In addition to this, however, the results presented so far raise further questions concerning the enzymes individually.

The molecular weight of the enzyme described by Louis and Fitt is remarkably low, as, indeed, are those of all the nucleic acid enzymes these authors have studied (Louis *et al.*, 1971). Among the functions a DNA-dependent RNA polymerase must possess is the ability to recognize specific promoter sequences on the DNA, and to respond to appropriate regulatory factors. The complexity, and hence the size, of the polymerase would therefore be expected to reflect the genomic complexity of the source (Küntzel and Schäfer, 1971). This is the case for the polymerase from the mitochondria of Neurospora crassa (64,000 daltons; Küntzel and Schäfer, 1971) and the phage-specific enzymes from T3 and T7 (110,000 daltons; Dunn *et al.*, 1971), which are smaller and respond to more restricted ranges of templates than the polymerase from E. coli (500,000 daltons; Burgess, 1971). It is remarkable that H. cutirubrum, with a genome size, and probably a genome complexity, similar to that of E. coli, would possess an RNA polymerase only a fraction of the size of the E. coli enzyme, and that the specificity of transcription by the two enzymes would be similar (Louis and Fitt, 1972a).

Although the size of the enzyme described by Chazan and Bayley (1973) is more reasonable, all attempts to obtain transcription of added template in concentrated salt failed. The most likely explanation is that in the native DNA-protein-membrane complex, the DNA is held in a form, possibly by the membrane itself, which facilitates the separation of the strands by the polymerase during transcription, and that this structural arrangement has not been reproduced with the isolated enzyme.

It is clear that further work is necessary before these divergent and somewhat contradictory results on halophilic RNA polymerases can be explained.

TRANSLATION

Studies on cell-free systems in my laboratory (Bayley and Griffiths, 1968a; Griffiths and Bayley, 1969) have shown that translation in H. cutirubrum is the same as in nonhalophilic bacteria in most respects, except most notably in being salt-dependent. As indicated in Figure 1, the *in vitro* incorporation of labeled amino acids into hot TCA-insoluble material requires saturated levels of monovalent cations for maximum activity. The ionic requirements are specific and the optimum mixture contains 3.8 M KCl, 1.0 M NaCl, and 0.4 M NH$_4$Cl as well as 0.04 M Mg acetate

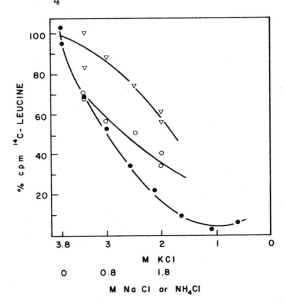

Fig. 1. Incorporation of [^{14}C]-leucine by a cell-free system from Halobacterium cutirubrum as a function of monovalent cation concentration. Potassium chloride concentration lowered with no compensating addition of other cations (—●—); KCl concentration lowered and compensated with added NaCl (—▽—) or NH$_4$Cl (—○—). In addition to the salts indicated, each incubation mixture contained 1.1 M NH$_4$Cl and 0.04 M Mg acetate. [Redrawn from Bayley and Griffiths (1968a)]

That this salt dependence is not due to the tRNAs
was demonstrated by work with heterologous systems from H.
cutirubrum and E. coli (White and Bayley, 1972c), in which
tRNAs were tested in concentrations of salt quite different
from those of their native environments. Transfer RNA
aminoacylated by its homologous synthetase transferred amino
acids into polypeptides, using heterologous ribosomes and
transfer factors. Furthermore, for several amino acids,
tRNAs from both bacteria were aminoacylated by heterologous
synthetases (Table 1). The recognition of a tRNA by its
cognate aminoacyl-tRNA synthetase is more precise than the
recognition involved in the transfer of amino acids on the
ribosome. The extent of heterologous interaction shown in
Table 1 is comparable to that found in other heterologous
systems involving nonhalophiles (e.g., Doctor and Mudd, 1963).

It is clear from these results, therefore, that the
structures of halophilic and nonhalophilic tRNAs cannot be
radically different, and that whatever configurational changes
are caused by changes in the salt concentration of the
environment must be reversible and similar in the tRNAs from
both bacteria.

The dependence of translation on salt in the H.cutirubrum
system is due to the protein components, concentrated salt
evidently being necessary for preserving functional
configurations of the proteins themselves, or for facilitating
the correct interactions of proteins with RNA, or for both.
With the aminoacyl-tRNA synthetases, activity falls as the
concentration of salt is reduced, in a similar fashion to
that for the whole amino acid incorporating system shown in
Figure 1 (Griffiths and Bayley, 1969). Experience in my
laboratory has shown that, although for optimal activity, all
the synthetases examined required 3.8 M KCl, requirements for
other monovalent cations varied quite widely.

Additional requirements for specific monovalent cations
in the complete system appear to be caused by the ribosomes
and polyribosomes. From studies on the sedimentation of
polyribosomes through density gradients under different
ionic conditions, Rauser and Bayley (1968) concluded that
NH_4^+ ions are essential to the binding of the nascent
polypeptide to the 50S ribosomal subunit.

The ribosomes of H. cutirubrum contain a preponderance
of acidic proteins, many of which remain bound to the rRNA
only in the presence of K^+ ions (Bayley and Kushner, 1964;

TABLE 1

Heterologous Interactions between tRNAs and Aminoacyl-tRNA Synthetases from H. cutirubrum and E. coli[a]

pmoles of ^{14}C-amino acid accepted per A_{260nm} unit of tRNA

^{14}C-labeled amino acid	E. coli tRNA		H. cutirubrum tRNA	
	E. coli Synthetases	H. cutirubrum Synthetases	H. cutirubrum Synthetases	E. coli Synthetases
Ala			14.6	5.3
Arg	9.8	21.1	12.5	2.5
Asp			8.6	1.3
Gly			10.3	2.6
His			4.1	1.9
Ile	8.8	8.4	9.6	1.2
Lys			4.8	3.8
Met	30.2	6.3	4.9	4.6
Ser	28.8	11.8		
Thr			4.6	13.8
Val	15.7	4.5	10.8	7.5

[a] From White and Bayley (1972c). Only significant heterologous interactions are shown, with the results for homologous systems for comparison.

Bayley, 1966a). As the concentration of K^+ ions is reduced, these acidic proteins dissociate, but reassociate to form 70S, 50S, and 30S particles when the ion concentration is restored (Bayley, 1966b). In the presence of high concentrations of monovalent cations other than K^+, the ribosomal subunits aggregate nonspecifically, as shown in Figure 2 (Bayley and Kushner, 1964).

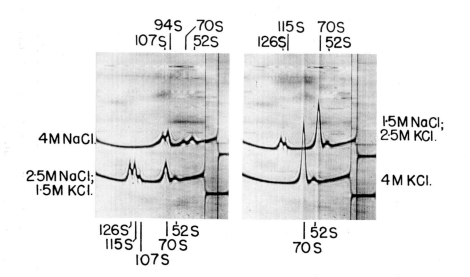

Fig. 2.   Sedimentation patterns of H. cutirubrum ribosomes, prepared in buffers containing 0.1 M MgCl₂, 0.01 M Tris-HCl, and the concentrations of monovalent salts indicated, to show nonspecific aggregation, except in the presence of 4 M KCl. [From Bayley and Kushner (1964).]

In all of these studies of the translational machinery of $H.$ cutirubrum, Mg^{2+} ions are required in concentrations very little higher than in nonhalophilic organisms, and they appear to play a similar role, namely, predominantly that of stabilizing RNA-RNA interactions both inter- and intra-molecularly (Bayley, 1966a; Griffiths and Bayley, 1969; Rauser and Bayley, 1968).

An extensive study of the ribosomes from $H.$ cutirubrum is underway by Matheson, Visentin, and co-workers (Visentin et al., 1972; Chow et al., 1972; Strøm and Visentin, 1973). This is part of a larger project to understand more of the essential features of ribosomes by comparing those from two bacteria living under extreme environmental conditions ($H.$ cutirubrum and Bacillus stearothermophilus) with those of $E.$ coli. The ribosomes from $H.$ cutirubrum and $E.$ coli are similar in many features, such as the size and composition of the subunits, the size and base composition of the RNA, and the numbers of different proteins; although the proteins from the halophile, in addition to being predominantly acidic, also range toward higher molecular weights. The total of the molecular weights of all proteins derived from preparations of the 30S subunit exceeds the amount of protein found in these subunits. This suggests that, as in $E.$ coli, the smaller subunit is structurally heterogeneous (Strøm and Visentin, 1973). Recent studies have established some homology in amino acid sequence between an acidic, alanine-rich protein from the 50S halophile subunit and the 50S ribosomal protein L7/L12 of $E.$ coli, which is involved in G-factor-dependent hydrolysis of GTP (Visentin, pers. commun.).

READING OF THE GENETIC CODE

In view of the extremely concentrated salt environment in which translation in $H.$ cutirubrum occurs, it is legitimate to inquire into the effect this salt may have on the reading of the genetic code. A further justification for such a study arises from the observation that many halophilic proteins, including those of the envelope (Brown, 1963; Kushner et al., 1964) and of the ribosome, are acidic (Reistad, 1970). If negative charges are an important feature of these proteins, enabling them to function in concentrated salt, then one possible explanation for the way extremely halophilic bacteria evolved would be that concentrated salt had led to a systematic misreading of codons for basic amino acids by tRNAs carrying acidic amino acids.

127

Many of the codons for basic and acidic residues differ by a single base in the 5' position. Thus, misreading could be achieved quite simply if salt induced wobble in this position during the matching of appropriate codons and anticodons (Bayley, 1966a).

Codon assignments have been investigated in the H. cutirubrum system using random heteropolyribonucleotides as messengers to direct both the incorporation of amino acids into polypeptides and the binding of aminoacyl-tRNAs to ribosomes (Bayley and Griffiths, 1968b; White and Bayley, 1972a). These experiments established the composition, but, of necessity, not the sequence, of codons assigned to different amino acids. The results are shown in Table 2 The relatively poor response of our crude halophile systems compared to the results achieved with E. coli has prevented our using polynucleotides of defined sequence as messengers in the incorporation experiments, and triplets of defined sequence in the binding assays, to establish codon sequences, and to extend the results to other codons, including those for acidic and basic amino acids.

One way of circumventing this problem is to improve the response in the binding assay by using purified tRNA species. Mr. J. Patel, in my laboratory (Patel, 1974), has obtained two species of lysine tRNA 90% pure and has evidence that the major species, representing 60% of the total tRNA^{Lys} in H. cutirubrum, responds to the codons AAA and AAG; whereas the minor species, representing the remaining 40% of lysine-accepting activity, responds to AAA, but only poorly, if at all, to AAG. Similar readings of these codons have been reported for tRNA^{Lys} species in B. subtilis (Chuang and Doi, 1972) and in eukaryotes, e.g., baker's yeast (Sen and Ghosh, 1973).

The results shown in Table 2 suggest that there is no gross change in codon assignments from the accepted code. Furthermore, studies on the fidelity of translation carried out with poly (U) (Bayley and Griffiths, 1968b), showed that salts affected fidelity only as their concentrations departed from the optimal 3.8 M KCl, 1.0 M NaCl, 0.4 M NH$_4$Cl, and 0.04 M Mg acetate. Under optimal conditions, leucine incorporation was 0.6% that of phenylalanine. This increased to 1% when the Mg acetate concentration was raised to 0.08 M. If, instead, the NH$_4$Cl concentration was increased to 1.1 M and the NaCl omitted, it became 2%. These results, together with studies on the codon specificity of binding of

TABLE 2

Possible Codon Assignments in H. cutirubrum *Based on Experiments with Synthetic Random Heteropolyribonucleotides*[a]

First Letter	Second Letter				Third Letter
	U	C	A	G	
U	Phe	Ser	Tyr	Cys	U
	Phe	Ser			C
	Leu				A
	Leu			Trp	G
C	Leu	Pro			U
	Leu	Pro	His	Arg	C
		Pro			A
		Pro		Arg	G
A	Ile				U
		Thr			C
	Ile		Lys[b]		A
			Lys[b]		G
G	Val			Gly	U
		Ala		Gly	C
					A
	Val (Met)	Ala			G

a
From Bayley and Griffiths (1968b); White and Bayley (1972a).

b
Based on a ribosomal binding assay using purified tRNA[Lys] species (Patel, 1974); not tested with other aminoacyl-tRNAs.

aminoacyl-tRNAs to heterologous ribosomes in the mixed $\underline{H}$. cutirubrum - $\underline{E}$. coli systems (White and Bayley, 1972a,c), all suggest that codon-anticodon recognition is unchanged in the halophile system.

[On this basis, the explanation for acidic proteins just suggested is incorrect. In any case, however, more recent results make the significance of negatively charged groups in halophilic proteins unclear, as not all halophilic enzymes are extremely acidic: the RNA polymerase of Louis and Fitt (1972a) is an example. A more important characteristic may be their polarity. Calculations of the hydrophobicity of halophilic proteins (Griffiths and Bayley, unpublished manuscript; Lanyi, 1974) based on the procedure of Bigelow (1967), show that they are significantly more polar than non-halophilic proteins. This increased polarity may be necessary to counteract the increased hydrophobic interactions created by "salting-out" salts like KCl in their environment (Lanyi, 1974).]

INITIATION OF PROTEIN SYNTHESIS

In studies on codon assignments, it was found that poly (GU) stimulated the incorporation of methionine by the $\underline{H}$. cutirubrum system (Bayley and Griffiths, 1968b). This suggested that GUG may be an initiator codon, as has been known for some time in the *in vitro* $\underline{E}$. coli system (Ghosh *et al.*, 1967) and has been shown *in vivo* in, for example, the RNA bacteriophage MS2 (Volckaert and Fiers, 1973). Initiation in organelles and cells containing 70S-type ribosomes appears to be brought about by N-formylmethionine-tRNA$_f^{Met}$, involving a distinctive initiator tRNA. In eukaryotic cells, with 80S-ribosomes, initiation is similar, except that formylation of methionine does not occur (Lucas-Lenard and Lipmann, 1971).

Attempts to detect formylation of methionyl-tRNA^{Met} in $\underline{H}$. cutirubrum, both *in vitro* and *in vivo*, were not successful (White and Bayley, 1972b). However, chromatography of crude tRNA charged with methionine on BD cellulose revealed two tRNA^{Met} species, one of which eluted with salt and the other with ethanol (White and Bayley, 1972b). The latter peak was formylatable in the $\underline{E}$. coli transformylation system (White and Bayley, 1972b). More recent experiments have shown that the ethanol species represents about 65% and the salt peak 35% of the total methionine-accepting activity (Dingle, 1973). Furthermore, it is the ethanol species, which, in addition to being recognized by the E. coli formylating enzyme, is the one recognized by the E. coli methionyl-tRNA synthetase to give the result shown in Table I. No aminoacylation of the salt species occurred in the heterologous system (Dingle, 1973).

Attempts to demonstrate unequivocally that the ethanol tRNA^{Met} species is an initiator in H. cutirubrum have unfortunately proved unsuccessful. However, the circumstantial evidence suggests that in contrast to most other 70S ribosomal systems, H. cutirubrum may resemble eukaryotic 80S ribosomal systems in initiating protein synthesis with nonformylated met-tRNA^{Met}. Work is in progress with Dr. RajBhandary to determine whether the putative initiator tRNA^{Met} itself resembles eukaryotic initiators in lacking the T - ψ - C sequence (Simsek et al., 1973).

IONIC INTERACTIONS IN INFORMATION TRANSFER

From the point of view of molecular biology, one of the attractive features of the extremely halophilic bacteria is that they offer an unusual system in which to study the importance of ionic interactions between biological macromolecules. The role such interactions play in E. coli has been clearly demonstrated in, for example, the binding of the lac repressor to the operator site (Riggs et al., 1970), and the binding of tRNA to the aminoacyl-tRNA synthetase (Loftfield and Eigner, 1967), although ionic interactions are probably significant in the binding of all proteins to nucleic acids.

Because of the high ionic strength within halobacteria, interactions between oppositely charged groups of the type encountered in nonhalophilic systems might not be expected. Nevertheless, Loftfield (1971, 1972) has shown that, despite the very concentrated salt involved, the activity of ile- and leu-tRNA synthetases from H. cutirubrum, measured by Griffiths and Bayley (1969), are linearly related to ionic activity as calculated by the Debye-Hückel theory. Loftfield interprets this relationship as suggesting that the ionic interactions between halophilic tRNA and its cognate synthetase are so tight that high ionic strengths are required to facilitate its release after aminoacylation.

To permit interactions between halophilic RNA and proteins to be examined more precisely, Mr. A. Pater has, in my laboratory, recently purified tryptophanyl-tRNA synthetase from H.

131

cutirubrum (Pater *et al.*, 1974). As is evident from other
contributions to this Symposium, the problem of purifying
halophilic proteins is that many methods require low ionic
strengths,under which conditions halophilic enzymes are usually
inactivated and often irreversibly denatured. We have
therefore resorted to affinity chromatography by adapting the
method devised by Robert-Gero and Waller (1972) for purifying
methionyl-tRNA synthetase from E. coli. Tryptophan is
attached through its carboxyl group to a support of Sepharose
4B by means of a "spacer" of hexamethylenediamine. A column
of this material effectively retains all of the tryptophanyl-
tRNA synthetase from a crude cell extract in the presence of
a buffer containing 3 *M* KCl. The synthetase can then be
eluted with a yield of about 40% and a purification of over
300-fold by adding L-tryptophan to the buffer (Figure 3).
This separation is much more effective than the separations
often obtained with nonhalophilic enzymes. One reason for
this is that the high ionic strength of the buffer eliminates
ion exchange effects on the column.

This method should provide us with enough material for
studies on the structure of the enzyme, as well as on its
mode of action.

CONCLUSION

Although extensive, detailed chemical analyses have not
been carried out on the nucleic acids of extremely halophilic
bacteria, it seems likely that these molecules function in
the same way as nonhalophilic nucleic acids in information
transfer; that is, the mechanism by which nucleic acids
recognize one another through Watson-Crick base pairing, and
possibly the features by which nucleic acids are recognized
by proteins, are probably unaffected directly by concentrated
salt. The more interesting question, however, is the way in
which the halophilic proteins involved in information transfer
have adapted to an environment of concentrated salt, and, in
particular, how these proteins interact with nucleic acids.
Some progress can now be hoped for in this area, and this
should prove valuable not only for a better understanding of
halophilic bacteria and of the way they have adapted to their
environment, but also for a better understanding of the
processes of information transfer themselves.

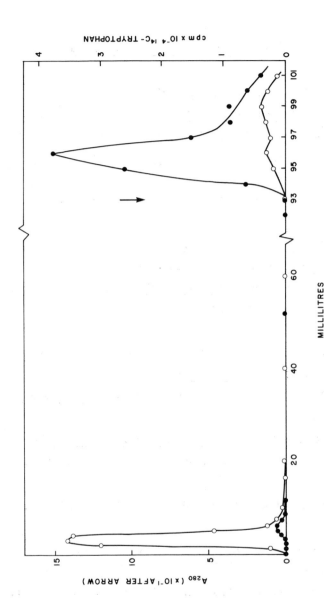

Fig. 3. The separation of tryptophanyl-tRNA synthetase from *H. cutirubrum* by affinity chromatography. A 10-ml column of substituted Sepharose 4B containing 5.8 μmole/ml tryptophan was equilibrated with buffer containing 3 M KCl, 0.1 M Mg acetate, 0.01 M Tris-HCl pH 8.05, and 8mM β-mercaptoethanol. Twelve milliliters of a dialyzed 150,000 g supernatant fraction of cell extract in the same buffer was applied and the column was eluted with this buffer, 20 mM L-tryptophan being added at the arrow. A_{280} — ○—; cpm ^{14}C-tryptophan esterified to tRNA, in test for synthetase activity —●—. Note the expansion of the abscissa and the A_{280} ordinate to the right of the arrow.

REFERENCES

ABRAM, D., and GIBBONS, N.E. (1960) Can. J. Microbiol. 6, 535.

BAYLEY, S.T. (1966a) J. Mol. Biol. 15, 420.

BAYLEY, S.T. (1966b) J. Mol. Biol. 18, 330.

BAYLEY, S.T., and GRIFFITHS, E. (1968a) Biochemistry, 7, 2249.

BAYLEY, S.T., and GRIFFITHS, E. (1968b) Can. J. Biochem. 46, 937.

BAYLEY, S.T., and KUSHNER, D.J. (1964) J. Mol. Biol. 9, 654.

BIGELOW, C.C. (1967) J. Theoret. Biol. 16, 187.

BROWN, A.D. (1963) Biochim. Biophys. Acta 75, 425.

BURGESS, R.R. (1971) Ann. Rev. Biochem. 40, 711.

CHAZAN, L.L., and BAYLEY, S.T. (1973) Can. J. Biochem. 51, 1297.

CHOW, C.T., VISENTIN, L.P., MATHESON, A.T., and YAGUCHI, M. (1972) Biochim. Biophys. Acta 287, 270.

CHRISTIAN, J.H.B., and WALTHO, J.A. (1962) Biochim. Biophys. Acta 65, 506.

CHUANG, R.Y., and DOI, R.H. (1972) J. Biol. Chem. 247, 3476.

DINGLE, C. (1973) M.Sc. Thesis, McMaster University, Hamilton, Ontario, Canada.

DOCTOR, B.P., and MUDD, J.A. (1963) J. Biol. Chem. 238, 3677.

DUNN, J.J., BAUTZ, F.A., and BAUTZ, E.K.F. (1971) Nature New Biol. 230, 94.

GHOSH, H.P., SÖLL, D., and KHORANA, H.G. (1967) J. Mol. Biol. 25, 275.

GRIFFITHS, E., and BAYLEY, S.T. (1969) Biochemistry, 8, 541.

JOSHI, J.G., GUILD, W.R., and HANDLER, P. (1963) J. Mol. Biol.
    6, 34.

KÜNTZEL, H., and SCHÄFER, K.P. (1971) Nature New Biol. 231,
    265.

KUSHNER, D.J., BAYLEY, S.T., BORING, J., KATES M., and
    GIBBONS, N.E. (1964) Can. J. Microbiol. 10, 483.

LANYI, J.K. (1974) Bact. Rev. 38, 272.

LARSEN, H. (1967) In "Advances in Microbial Physiology"
    (A.H. Rose and J.F. Wilkinson, eds.), Vol. 1, 97,
    Academic Press, New York.

LOFTFIELD, R.B. (1971) In "Protein Synthesis" (E. McConkey,
    ed.), Vol. 1, 1, Marcel Dekker, New York.

LOFTFIELD, R.B. (1972) Progr. Nucleic Acid Res. 12, 87.

LOFTFIELD, R.B., and EIGNER, E.A. (1967) J. Biol. Chem. 242,
    5355.

LOU, P.L. (1970) M. Sc. Thesis, McMaster University, Hamilton,
    Ontario, Canada.

LOUIS, B.G., and FITT, P.S. (1971a) Biochem. J. 121, 621.

LOUIS, B.G., and FITT, P.S. (1971b) FEBS Letters, 14, 143.

LOUIS, B.G., and FITT, P.S. (1972a) Biochem. J. 127, 69.

LOUIS, B.G., and FITT, P.S. (1972b) Biochem. J. 127, 81.

LOUIS, B.G., PETERKIN, P.I., and FITT, P.S. (1971) Biochem.
    J. 121, 635.

LUCAS-LENARD, J., and LIPMANN, F. (1971) Ann. Rev. Biochem.
    40, 409.

MOORE, R.L., and MCCARTHY, B.J. (1969a) J. Bacteriol. 99,
    248.

MOORE, R.L., and MCCARTHY, B.J. (1969b) J. Bacteriol. 99,
    255.

PATEL, J. (1974) M.Sc. Thesis, McMaster University, Hamilton,
    Ontario, Canada.

PATER, A., BAYLEY, S.T., and CHAN, W.W. (1974) Fed. Proc. 33, 1313A.

RAUSER, W.E., and BAYLEY, S.T. (1968) J. Bacteriol. 96, 1304.

REISTAD, R., (1970) Arch. Mikrobiol. 71, 353.

RIGGS, A.D., BOURGEOIS, S., and COHN, M. (1970) J. Mol. Biol. 53, 401.

ROBERT-GERO, M., and WALLER, J.P. (1972) Eur. J. Biochem. 31, 315.

SEN, G.C., and GHOSH, H.P. (1973) Biochim. Biophys. Acta 308, 106.

SIMSEK, M., ZIEGENMEYER, J., HECKMAN, J., and RAJBHANDARY, U.L. (1973) Proc. Nat. Acad. Sci. U.S. 70, 1041.

SINGER, C.E., and AMES, B.N. (1970) Science, 170, 822.

STRØM, A.R., and VISENTIN, L.P. (1973) FEBS Letters, 37, 274.

VISENTIN, L.P., CHOW, C., MATHESON, A.T., YAGUCHI, M., and ROLLIN, F. (1972) Biochem. J. 130, 103.

VOLCKAERT, G., and FIERS, W. (1973) FEBS Letters, 35, 91.

WHITE, B.N., and BAYLEY, S.T. (1972a) Can. J. Biochem. 50, 600.

WHITE, B.N., and BAYLEY, S.T. (1972b) Biochim. Biophys. Acta 272, 583.

WHITE, B.N., and BAYLEY, S.T. (1972c) Biochim. Biophys. Acta 272, 588.

SALT-MEDIATED KILL AND INHIBITION OF THERMALLY INJURED (RIBOSOME DEFICIENT) *STAPHYLOCOCCUS AUREUS*

Roger D. Haight

Tomlins *et al.* (1971) have indicated that thermal injury
causes numerous biochemical lesions in bacterial cells.   In
Staphylococcus aureus these lesions include enzyme damage
(glycolysis and TCA cycle), loss of ribosomal RNA, a decrease
in the internal K^+/Na^+ ratio, a loss of salt (NaCl) tolerance
for growth, a loss of membrane lipids, and the creation of
an extended lag period for growth onset. Recovery from sub-
lethal thermal injury occurred during the extended lag period
for injured cells, and this process was uniquely marked by a
parallel recovery of growth tolerance to 7.5% salt. By the
judicious use of inhibitors, Iandolo *et al.* (1966) showed
that salt tolerance recovery did not require cell division,
cellwall synthesis, or protein synthesis. Hurst *et al.* (1973)
showed that membrane lipid resynthesis was not required for
salt tolerance recovery. Ribosomal RNA synthesis had been
clearly shown by Sogin and Ordal (1967) to parallel the return of salt tolerance, and that RNA synthesis inhibition prevented the recovery of salt tolerance by injured cells.

Virtually all of these studies were performed using a
double-medium technique, wherein salt was used to distinguish
the salt-sensitive from the salt-insensitive portions of the
viable cell population remaining after thermal stressing.

137

On a Trypticase Soy Agar medium containing 7.5% salt (TSAS), uninjured or recovered cells formed colonies, whereas, on the same medium without salt (TSA), injured (salt-sensitive), uninjured, and recovered cells (salt-tolerant) formed colonies equally well. Using the lack of colony formation on TSAS as an index of unrepaired injury, recovery from the injury could then be followed.

The lack of colony formation on TSAS--the manifestation of salt intolerance--could have been due to a lethal or inhibiting action of salt or one of its ions, or to an osmotic effect on the injured cells. The data presented deal primarily with distinguishing the mode of action of salt on injured cells.

Cells of Staphylococcus aureus MF-31, normally salt tolerant, were grown in Trypticase Soy Broth (TSB) for 16-18 hours at 37°C, with shaking. Following harvest and washing of a 20-ml portion of the broth culture, the cell pellet was resuspended in 25 ml of 100 mM pH 7.2 potassium phosphate buffer (PPB). A 10-ml portion of the resultant PPB cell suspension was then added to 90 ml of preheated (55°C) PPB, with mixing. This suspension was then continuously heated at 55°C, with shaking, for 10 minutes. This procedure injured the cells, rendering them salt sensitive. For the recovery of salt tolerance, a 10-ml portion of the injured PPB cell suspension was transferred into 90 ml of 1.11×TSB (37°C) or other storage media, and incubated. Samples were removed on timed intervals and were diluted in sterile distilled water. Dilution samples were then surface-plated onto TSA and TSAS. The resultant colonies were counted after incubation at 37°C for 48 hours. Figure 1 shows typical results obtained from this procedure. The open circles show a lag of 4 hours in the growth pattern of injured cells incubated (stored) in TSB, as detected on TSA plates. The filled circles show a rapid increase in the salt-tolerant population (TSAS counts) that resulted from incubation in TSB. Note that nearly all of the cells regained salt tolerance before the onset of cell division in TSB (TSA counts, -o-), and that the injured cells, at the time of addition to TSB, were salt-sensitive (difference in counts between TSA and TSAS at zero storage time).

When the storage broth contained salt (TSBS), the broth equivalent of TSAS in its effect, stored injured cells exhibited a rapid decline in expressed viability on TSA (Figure 1, -Δ-), the medium that normally supported colony formation of injured cells. This decline was complete within 3 hours,

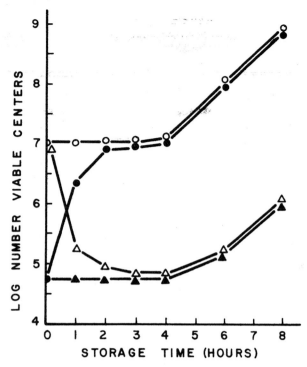

Fig. 1. *Effect of storage in TSB and TSBS on thermally
stressed cells. Cells were rendered salt-sensitive
by heating for 10 minutes at 55°C in 100 mM phosphate buffer (pH 7.2). One part of heated cell suspension was added to nine parts of TSB or TSBS at
37°C. Samples were removed on timed intervals, and
enumerated on TSA and TSAS. Symbols: TSA counts of
cells stored in TSB (○), TSAS counts of cells stored
in TSB (●), TSA counts of cells stored in TSBS (△),
TSAS counts of cells stored in TSBS (▲). Erwin and
Haight (1973).*

and was interpreted as salt-effected death of salt-sensitive
cells. Injured cells continuously stored in TSBS and plated
onto TSAS showed a typical 4-hour lag before the onset of
cell division in TSBS.

The presumption from these data was that all salt-sensitive cells in an injured population would be killed by salt,

until that point in the recovery process at which full sal⁺ tolerance returned; that is, that no immunity to salt kill would develop during recovery, until the return of full salt-tolerance. This presumption was tested by suspending injured (salt-kill sensitive) cells in TSB for varying periods of time, from zero through 3 hours. At the end of the desired TSB storage period, the cells were transferred into TSBS for a 2-hour storage period. The TSB storage periods would be conducive to salt-tolerance recovery, whereas the TSBS storage period should kill all salt-sensitive (unrecovered) cells. Thus, the cells stored only in TSB should yield typical TSA plate count values, whereas those samples secondarily stored in TSBS should yield TSA counts equivalent to TSAS counts, for TSB-only-stored cells. In the event that TSB-stored recovering cells did not recover full salt tolerance in a single step, but rather recovered full salt tolerance in a stepwise manner involving a stage of immunity to salt-kill, while retaining sequentially a salt-inhibition characteristic, then the sequentially TSB-stored samples should have yielded TSA plate counts greater than TSB-stored cells plated on TSAS. Figure 2 shows the data obtained from these experiments. The filled squares show the typical recovery curve for TSB-stored cells plated on TSAS (unrecovered cells in the population were either killed or inhibited on

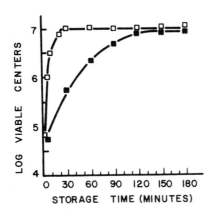

Fig. 2. Effect of stressed cell storage in TSB, PPB, or PPBS preceding subjection to TSBS. Cells were rendered salt-sensitive as in Figure 1. Fractions of stored cells were transferred on timed intervals into TSBS (37°C) for 120 minutes prior to enumeration on TSA and TSAS. Symbols: TSB-, PPB-, PPBS-stored cells enumerated on TSA after 120-min subjection to TSBS (□); TSB-, PPB-, PPBS-stored cells enumerated on TSAS after 120-min subjection to TSBS (■). Erwin and Haight (1973).

140

the TSAS medium). The open squares show the data obtained for TSB-stored cells (for the times indicated) followed by a 2-hour TSBS storage prior to plating on TSA. The data clearly indicated that recovering cells regain full salt-tolerance in a stepwise sequence from killable by salt, to inhibited by salt (immune to salt kill), to full salt tolerance.

The question of what ion, ions, or property of salt exerted the lethal and inhibitory effect was examined by fortifying TSB with glycerol of an equivalent osmotic pressure to 7.5% NaCl (TSBO), and by preparing TSB containing an equimolar amount of KCl to that of 7.5% NaCl (TSBK). When these broths were used as the storage menstruum, the data shown in Figure 3 were obtained. The open circles represent the plate counts for injured cells stored in TSB, TSBO, and TSBK, enumerated on TSA: the standard viability system. The filled circles show the TSAS plate counts obtained for injured cell storage in TSB or TSBO. In this case, the cells recovered salt tolerance normally, or slightly better, as compared to the open triangle curve, which represents the recovery data for TSBK-stored, TSAS-enumerated, injured cells. Thus, it appeared from these data that Cl⁻ ion and osmotic pressure

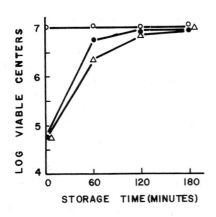

Fig. 3. Salt tolerance recovery of thermally stressed cells stored in TSBK or TSBO. Cells were rendered salt-sensitive as in Figure 1. One part of cell suspension was added to nine parts of storage medium at 37°C. Samples were removed on timed intervals and enumerated on TSA and TSAS. Symbols: TSA count of cells stored in TSB, TSBK, or TSBO (○); TSAS count of cells stored in TSB or TSBO (●); TSAS count of cells stored in TSBK (Δ). Erwin and Haight (1973).

were not involved in the NaCl lethal and inhibitory effects
observed previously.

Since injured cells stored in nutrient media were metabolically active, the question of a metabolic linkage to salt-
effected kill was raised. This potential linkage was examined
by storing salt-sensitive cells in PPB and PPBS for various
time periods before plating the cells on TSA and TSAS.
Figure 4 (open circles) shows the typical TSB-stored, TSA-
enumerated viability system, and the closed circles show the
typical TSB-stored, TSAS-enumerated recovery curve. The open
triangles show PPB or PPBS-stored cells enumerated on TSA,
and the closed triangles show PPB or PPBS-stored cells enumerated on TSAS. The data indicated that injured cells did not
recover salt-tolerance, and in fact, lost viability slowly
when stored either in PPB or PPBS. The lack of salt-tolerance
recovery in PPB or PPBS was expected, since Sogin and Ordal
(1967) have shown that ribosomal RNA resynthesis (rRNA was
destroyed during injury process) was required for salt-tolerance recovery. The slow viability loss observed for cells
stored in PPBS was not nearly as rapid as that observed for
TSBS-stored cells, indicating that active nutrient metabolism
was required for salt-kill to be exerted. The nature of this

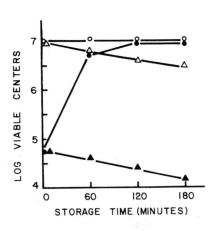

STORAGE TIME (MINUTES)

*Fig. 4. Effect of
stressed cell storage in TSB,
PPB, and PPBS. Cells were
rendered salt-sensitive as
in Figure 1. One part of
cell suspension was added to
nine parts of storage medium
at 37°C. Samples were removed on timed intervals and
enumerated on TSA and TSAS.
Symbols: TSA-enumerated
cells stored in TSB (○);
TSA-enumerated cells stored
in PPB or PPBS (△); TSAS-
enumerated cells stored in
TSB (●); TSAS-enumerated
cells stored in PPB or PPBS
(▲). Erwin and Haight
(1973).*

linkage has not been elucidated, and its potential relationship to the metabolic dysfunction of injured cells, noted by
Bluhm and Ordal (1969), is unknown.

Finally, the metabolic linkage to the development of
immunity to salt-kill (transition from killable by salt to
inhibited by salt), was examined by storing salt-sensitive
cells in PPB and PPBS for varying timed periods, prior to
transfer into TSBS (2-hour storage in nutrient media to kill
any salt-sensitive cells remaining). The data obtained for
these experiments were essentially the same as shown in
Figure 2 [-□-, PPB and PPBS-stored cells, transferred into
TSBS (2 hours), TSA enumerated; -■-, PPB and PPBS-stored
cells, transferred into TSBS (2 hours), TSAS enumerated],
clearly showing that immunity development was not metabolically linked, nor was it interfered with by salt, when
nutrients were absent. These data clearly suggested that the
development of immunity to salt-kill was occurring physically.
The mechanism of this process remains to be elucidated, but
it could have occurred via a membrane melting-reannealment
process similar to that proposed by Steim (1972).

In summary we have shown that thermal-injury of Staphylococcus aureus MF-31 rendered cells salt-sensitive, and that
the process of recovery of salt-tolerance which occurred in
the absence of cell division, involved a stepwise transition
from killable by salt, to inhibitable by salt, to full salt
tolerance. The first transition stage (killable to inhibitable by salt, called the development of immunity to salt kill)
occurred physically, and was apparently not metabolically
linked, whereas the salt-effected kill process did appear to
require a metabolic linkage. Furthermore, for kill and/or
inhibition, Cl^- ion or osmotic effect was shown to not be
involved; Na^+ ion was implicated as the active agent.

ACKNOWLEDGMENT

The author wishes to acknowledge Daniel G. Erwin, whose
effort in the laboratory made this paper possible. The
author also wishes to acknowledge the Research Corporation
(Brown-Hazen Fund) and the SJSU Foundation for financial
assistance.

REFERENCES

BLUHM, L., and ORDAL, Z. J. (1969). J. Bacteriol. 97, 140.
ERWIN, D. G., and HAIGHT, R. D. (1973). J. Bacteriol. 116, 337.

HURST, A., HUGHES, A., BEARE-ROGERS, Joyce L., and COLLINS-THOMPSON, D. L. (1973). J. Bacteriol. 116, 901.

IANDOLO, J. J., and ORDAL, Z. J. (1966). J. Bacteriol. 91, 134.

SOGIN, S. J., and ORDAL, Z. J. (1967). J. Bacteriol. 94, 1082.

STEIM, J. M. (1972). Second Symposium on "Extreme Environments: Mechanisms of Microbial Adaptation." National Aeronautics and Space Administration, Ames Research Center, Moffett Field, California.

TOMLINS, R. I., PIERSON, M. D., and ORDAL, Z. J. (1971). Canad. J. Microbiol. 17, 759.

PROTEINS

PROTEINS FROM THERMOPHILIC MICROORGANISMS

Lars G. Ljungdahl and David Sherod

Most life as we know it seems to have an upper temperature limit somewhere between 40° and 50°C. However, several
microorganisms live, grow, and reproduce at much higher temperatures. They are called thermophilic, to distinguish them
from mesophilic microorganisms, which grow optimally between
30° and 45°C, and the psychrophilic microorganisms, which
grow optimally below 20°C (Stanier et al., 1970). The existence of thermophilic organisms has triggered research and
speculation as to how these organisms evolved, and how they
are able to function at high temperatures. Since proteins,
especially enzymes, normally are denatured at temperatures
below the optimum growth temperatures of thermophiles, much
research has been concerned with proteins isolated from thermophiles. The properties of proteins from thermophilic
microorganisms, is the subject of this review.

Thermophilic organisms do not constitute a particular
genus or genera. Thus thermophilic organisms exist among
algae (Castenholz, 1969), fungi (Cooney and Emerson, 1964),
and several bacterial genera (Allen, 1953; Farrell and Campbell, 1969; Brock and Freeze, 1969). Taxonomically, thermophilic species are related to the mesophilic species. The
thermophile's gross anatomy, ultrastructure, respiration, and

metabolic processes are very similar, if not identical, when compared with mesophilic organisms. These similarities indicate that thermophiles and mesophiles have evolved from common ancestors. This becomes an important point when one considers that most of the work which has been done, and which is now being done, to find the molecular basis for thermophilicity, involves a comparison between structures, particles, and macromolecules isolated from thermophiles, with corresponding counterparts isolated from mesophiles. If thermophiles were evolved independently from mesophiles, such comparative work would be difficult and would probably be of little value.

If we accept the concept that thermophilic and mesophilic microorganisms have a common origin, we must accept that their structures, particles, and macromolecules evolved from a common source. We would then predict that an enzyme from a thermophilic organism would be similar to the same enzyme from a mesophilic organism, at least, if the two organisms are species of the same genus. This is, in general, what has been observed, except that proteins from thermophilic organisms often have substantially higher thermal stability than corresponding proteins from mesophilic organisms.

At this point, it is appropriate to point out that the so-called optimum temperature obtained by determining the velocities of an enzyme reaction at different temperatures, and by plotting these velocities against the temperature, is not necessarily a measure of thermostability. Thus, the optimum temperature of an enzyme reaction is complex and variable. It depends on how the velocity was measured (initial velocity or not) and also on the rate of denaturation of the enzyme. A protein's thermostability is better measured by studying its rate of denaturation, independently of its enzyme activity (Dixon and Webb, 1964).

STABILITY OF ENZYMES IN CELL FREE EXTRACTS

Militzer *et al.* (1949) were among the first to demonstrate that enzymes in extracts from thermophilic microorganisms are more heat stable than enzymes in extracts from mesophiles. They studied the activity of malate dehydrogenase in extracts from an obligate thermophilic bacterium, and found that the activity was stable for 120 min at 65°C. This was in contrast with the malate dehydrogenase activity in extracts from the mesophile Bacillus subtilis, which is rapidly inactivated at 65°C.

Howell *et al*. (1969) found that glycolytic enzymes and 6-phosphogluconate dehydrogenase, in crude extracts from the thermophiles Clostridium tartarivorum and Clostridium thermosaccharolyticum, are more heat stable than the corresponding enzymes from the mesophile Clostridium pasteurianum, with the exception of pyruvate kinase. However, the thermostability of this enzyme was found to be high in all three clostridia.

Clostridium thermoaceticum, a thermophile, and Clostridium formicoaceticum, a mesophile, ferment hexoses almost quantitatively to 3 moles of acetate, of which one is synthesized from CO_2 and pyruvate (Schulman *et al*., 1973). The synthesis of acetate from CO_2 and pyruvate involves a battery of enzymes, including those which catalyze conversion of one-carbon compounds of tetrahydrofolate (O'Brien and Ljungdahl, 1972; Andreesen *et al*., 1973). Cell free extracts of C. thermoaceticum catalyze the acetate synthesis at 57°C (Ljungdahl *et al*., 1965; Poston *et al*., 1966), but they are almost inactive below 35°C. This indicates that all the required enzymes for acetate synthesis in C. thermoaceticum are quite thermostable. Many enzymes required for acetate synthesis in C. formicoaceticum, on the other hand, are rapidly inactivated at temperatures above 45°C (Ljungdahl, unpublished).

Amelunxen and Lins (1968) compared the thermostability of 11 enzymes in extracts from the thermophile Bacillus stearothermophilus and the mesophile Bacillus cereus. They found that nine of the B. stearothermophilus enzymes are substantially more heat stable than corresponding enzymes from the mesophile.

It should also be mentioned that the complete protein-synthesizing machinery from B. stearothermophilus (Friedmann, 1968), as well as from two thermophilic clostridia (Irwin *et al*., 1973), has much higher thermal stability than that from mesophilic microorganisms.

These and similar studies show that in obligate thermophilic microorganisms, most enzymes are thermostable when present in crude extracts. However, these studies do not reveal the molecular basis for the higher thermostability. To answer this question it is necessary to obtain pure homogeneous enzymes.

GENERAL PROPERTIES OF PROTEINS FROM THERMOPHILES

Several enzymes and other proteins from thermophilic microorganisms have now been purified to homogeneity, and some of these are listed in Table 1. Since Singleton and Amelunxen (1973), in a recent review, discussed the properties of several of these enzymes and proteins, and compared their properties with counterparts isolated from mesophiles, we will not discuss these proteins in detail, but will merely summarize the results. First, it can be concluded that pure proteins from thermophilic microorganisms are, in general, more thermostable than their counterparts from mesophilic organisms. Furthermore, in all cases except one, proteins from thermophiles are remarkably similar in physical properties such as molecular weight, number of subunits, Stokes radius, specific volume, and sedimentation constant. Thus, in terms of physical parameters, proteins from thermophilic microorganisms do not differ much from their mesophilic counterparts.

The only thermostable protein found so far that differs grossly from its mesophilic counterpart is an α-amylase purified from B. stearothermophilus by Manning *et al.* (1961). This enzyme appears to be in a semirandom coil conformation and thus exists in a semidenatured state. It is stabilized by a disulfide bridge and has a very small molecular weight, compared with α-amylases from other bacteria, including strains of B. stearothermophilus (Isono, 1970; Ogasahara *et al.*, 1970; Pfueller and Elliot, 1969). This enzyme was one of the first isolated from thermophilic organisms, and it was subsequently thought that enzymes from thermophiles should be small, and exist in semirandom coil conformations. That enzymes from thermophiles should have such different properties, as compared with enzymes from mesophiles, is not only incompatible with the concept of evolution, but also with the experimental findings, which show that most enzymes isolated from thermophiles have physical properties very similar to enzymes from mesophiles.

NONPROTEIN FACTORS IN STABILIZATION OF PROTEINS

Theories to explain the high stability of proteins in thermophiles include stabilization by nonprotein factors. This seems unlikely since the purified proteins from thermophiles, listed in Table 1, possess little if any material except amino acids; if such a factor is present it must be tightly bound to the protein and constitute a very small part of the molecule. A thorough search for carbohydrates and lipids in formyltetrahydrofolate synthetase from C. thermoaceticum (Ljungdahl *et al.*, 1970) gave a negative result. Similarly the glyceraldehyde 3-phosphate dehydrogenase from

B. stearothermophilus does not contain any carbohydrate
(Singleton *et al.*, 1969). Barnes and Stellwagen (1973) concluded, after a thorough search, that enolase from Thermus
X-1 does not contain any nonprotein material. Amino acid
analyses have been performed on many of the thermostable proteins listed in Table 1, and in most cases the recoveries of
amino acids have been close to 100%, leaving little room for
the presence of other chemicals. However, thermolysin from
Bacillus thermoproteolyticus is stabilized by calcium ions
(Feder *et al.*, 1971), and Nakamura (1960) found a "factor S,"
which increased the thermostability of catalase from a thermophilic bacterium by about 5°C.

Proteins from thermophiles often have higher thermostability in the presence of substrate, metals, activators,
inhibitors, or other co-factors. However, this is also true
for proteins from mesophilic organisms. Thus, in the absence
of these factors the proteins from thermophiles have higher
thermostability than corresponding mesophilic proteins. In
fact, in the presence of substrate some thermophilic proteins
have lower heat stability (Howard and Becker, 1972).

A possible mechanism for increasing the thermostability
of proteins is structural stabilization in the form of macromolecular complexes. That is, proteins interact with each
other to form larger conjugates, which are more thermostable
than individual proteins. Some experimental support has been
obtained for this (Donovan and Beardslee, 1974).

STABILIZATION BY DISULFIDE AND HYDROPHOBIC BONDS

Almost all evidence so far accumulated regarding thermal
stability of proteins from thermophiles leads to the conclusion that these proteins have intrinsic thermostability,
which depends on their amino acid composition and the
sequence of the amino acids. Thus, in proteins from thermophiles, some amino acids which, due to their properties,
enhance the thermostability are substituted for amino acids
that do not strengthen the stability. Such substitutions
may add bonds that stabilize the secondary and tertiary
structure of proteins, as, for example, disulfide bridges;
hydrogen bonds; ionized group interactions; and hydrophobic
bonds. Heat effects and factors influencing thermal stability of proteins have been discussed at length by Brandts
(1967). See also Waugh (1954) for a more general discussion
of interactions between the main peptide chain and the side
chains in proteins.

Table 1

Example of Proteins Purified from Thermophilic Microorganisms

Enzyme	Microorganism	Reference
Ferredoxin	C. tartarivorum	Devanathan et al. (1969).
	C. thermosaccharolyticum	Devanathan et al. (1969).
Cytochrome[c]	H. lanuginosa	Morgan et al. (1972);
		Morgan and Riehm (1973).
Thermolysin	B. thermoproteolyticus	Feder et al. (1971).
Glucose-6-phosphate dehydrogenase	P. dupontii	Broad and Shepherd (1970).
Glucose-6-phosphate isomerase	B. stearothermophilus	Muramatsu and Nosoh (1971).
Glyceraldehyde-3-phosphate	B. stearothermophilus	Singleton et al. (1969);
dehydrogenase		Amelunxen et al. (1970);
		Bridgen et al. (1972);
		Amelunxen (1967).
Enolase	T. aquaticus YT-1	Stellwagen et al. (1973).
Aldolase	B. stearothermophilus	Thompson and Thompson (1962);
		Sugimoto and Nosoh (1971).
Isocitrate dehydrogenase	B. stearothermophilus	Howard and Becker (1970).
(TPN$^+$-dependent)		
Formyltetrahydrofolate synthetase	C. thermoaceticum	Ljungdahl et al. (1970).
Methylenetetrahydrofolate	C. thermoaceticum	O'Brien et al. (1973).
dehydrogenase		
Pyrimidine-nucleoside: ortho-	B. stearothermophilus	Saunders et al. (1969).
phosphate ribosyl transferase		
Pyrimidine ribonucleoside kinase	B. stearothermophilus	Orengo and Saunders (1972).
DNA-dependent RNA polymerase	B. stearothermophilus	Stenesh and Roe (1972).
Tyrosyl-tRNA synthetase	B. stearothermophilus	Koch (1974).

Table 1 (continued)

Example of Proteins Purified from Thermophilic Microorganisms

Enzyme	Microorganism	Reference
Valyl-tRNA synthetase	B. stearothermophilus	Wilkinson and Knowles (1974).
Ribosomal proteins	B. stearothermophilus	Friedmann (1968); Ansley et al. (1969); Saunders and Campbell (1966); Higo and Loertscher (1974); Yaguchi et al. (1973).
ATPase	B. stearothermophilus	Hachimori et al. (1970).
Alkaline proteinase	Streptomyces	Mizusawa and Yoshida (1972).
Protease	B. thermoproteolyticus	Ohta (1967).
Aminopeptidase I	B. stearothermophilus	Roncari and Zuber (1969); Stoll et al. (1972); Moser et al. (1970).
α-amylase	B. stearothermophilus	Isono (1970); Pfueller and Elliot (1969); Campbell and Manning (1961).

Disulfide bonds may stabilize proteins. Ribonuclease has
high thermal stability. It contains four disulfide bonds
(Spackman *et al.*, 1960; Smyth *et al.*, 1963), and it is likely
that they contribute to the thermal stability. However,
available data from thermophilic proteins does not show that
such bonds are more numerous in proteins from thermophilic
organisms. It appears, in fact, that there are less disulfide
bonds in proteins from thermophiles, as compared with proteins
from mesophiles. Table 2 lists the number of half-cystine
residues in some proteins from thermophiles and in their
counterparts from mesophiles. Ferredoxin has eight half-
cystine residues whether or not it is isolated from a thermophile or from a mesophile, and all eight residues are involved
in binding the iron atoms. Aminopeptidase I from B. stearothermophilus and a protease from B. thermoproteolyticus do not
have any half-cystine residues, and are consequently not stabilized by disulfide bridges. The thermostable enolase also
contains no half-cystine residues, while the less thermostable
enolase does. Thus, it becomes apparent that disulfide bonds
are not a necessity for thermal stability. Formyltetrahydrofolate synthetase has 24 half-cystine residues (six per subunit), but none of these forms disulfide bonds (Ljungdahl *et
al.*, 1970; Himes and Harmony, 1973). Furthermore, some proteins from thermophilic organisms actually are activated by
sulfhydryl compounds, for example, the extremely thermostable
aldolase from T. aquaticus (Freeze and Brock, 1970). These
and similar reports show that stabilization by disulfide
bridges is not a general mechanism for the thermal stability
of proteins from thermophiles. Of interest is the finding by
Mizusawa and Yoshida (1973). They reported that a —SH group
participates in a tight organization of the active site of
alkaline proteinase from the thermophile Streptomyces rectus
var. proteolyticus, and that this conferred high thermal
stability.

Hydrophobic bonds are very important to the stability of
proteins, and proteins may be stabilized almost completely by
them (Tanford, 1962; Bigelow, 1967). Furthermore, the stability of these bonds increases with temperature to about 65°C
(Brandts, 1967). Therefore, it has been postulated by a number of investigators that thermophilic proteins may have more
amino acids with hydrophobic side chains, than have the mesophilic proteins. A protein's hydrophobicity may be calculated
by three different methods: average hydrophobicity ($H\phi_{avg}$)
according to Bigelow (1967), the frequency of nonpolar side
chains (NPS) according to Waugh (1954), and the ratio (p) of
the volume occupied by polar residues to that occupied by

Table 2

Half-Cystine Residues in Proteins from Thermophilic and Mesophilic Organisms

Protein	Organism	Half-cystine (residues/mole)	Reference
Ferredoxin	C. tartarivorum[a]	8.0	Devanathan et al. (1969).
	C. thermosaccharolyticum[a]	8.0	Devanathan et al. (1969).
	C. pasteurianum	8.0	Tanaka et al. (1966).
Aminopeptidase I	B. stearothermophilus	0	Roncari and Zuber (1969).
Cytochrome c	Humicola lanuginosa[a]	2.0	Morgan et al. (1972).
	Neurospora crassa	2.0	Morgan et al. (1972).
Protease	B. thermoproteolyticus[a]	0	Ohta et al. (1966).
Tyrosyl-tRNA synthetase	B. stearothermophilus[a]	4.0	Koch (1974).
Glyceraldehyde-3-phosphate dehydrogenase	B. stearothermophilus[a]	12.0	Singleton et al. (1969).
	lobster muscle	20.0	Singleton et al. (1969).
	pig muscle	16.0	Singleton et al. (1969).
Enolase	Thermus X-1[a]	0	Barnes and Stellwagen (1973).
	T. aquaticus YT-1[a]	0	Barnes and Stellwagen (1973).
	rabbit muscle	6.0	Barnes and Stellwagen (1973).
	yeast	1.0	Barnes and Stellwagen (1973).
Methylenetetrahydrofolate dehydrogenase	C. thermoaceticum[a]	8.0	O'Brien et al. (1973).
	C. formicoaceticum	7.0	Moore et al. (1974).
Formyltetrahydrofolate synthetase	C. thermoaceticum[a]	24.0	Ljungdahl et al. (1970).
	C. cylindrosporum	24.0	Himes and Harmony (1973).
	C. acidiurici	24.0	Himes and Harmony (1973).

[a] *Thermophilic organisms.*

155

nonpolar residues, according to Fisher (1964). Hydrophobicity
has been calculated for 10 proteins from both mesophilic and
thermophilic organisms by Singleton and Amelunxen (1973), and
also for a large number of proteins by Bigelow (1967) and
Goldsack (1970). Some of the thermostable proteins have
slightly higher hydrophobicity as compared with the same proteins from mesophilic organisms. For example, the $H\phi_{avg}$ of
formyltetrahydrofolate synthetase from C. thermoaceticum is
1160 calories per amino acid residue, whereas the same enzyme
from C. formicoaceticum has a $H\phi_{avg}$ of 1120 calories per residue, and that of C. cylindrosporum 1050 calories per residue
(Ljungdahl et al., 1970). However, more thermostable proteins
are not more hydrophobic when compared to their counterparts
from mesophiles. Thus, glyceraldehyde 3-phosphate dehydrogenase from B. stearothermophilus has a $H\phi_{avg}$ of 1060 calories
per residue, which is low compared to the enzyme isolated from
other sources (1050-1160 calories per residue) (Singleton and
Amelunxen, 1973). Thus, the calculations of average hydrophobicity values for proteins have failed to support the idea
that thermostable proteins are more hydrophobic. This, of
course, does not mean that hydrophobic bonds are not involved
in the stability of thermostable proteins. It is important to
realize that rather small changes in the energy of stabilization are needed to confer a higher thermal stability to a
protein. In the example of formyltetrahydrofolate synthetase,
the $H\phi_{avg}$ difference between the C. thermoaceticum enzyme and
the C. formicoaceticum enzyme is 40 calories per residue.
There are approximately 530 amino acids per subunit, and it is
easily seen that small changes in average hydrophobicity will
make a substantial change in available energy for stabilization (Ljungdahl et al., 1970). It seems obvious from the preceding discussion, that we must look more closely into the
amino acid sequence and the role of single amino acids strate-
gically placed in that sequence, to explain thermal stability
of proteins. The proteins may form extra H-bonds, have ionized group interactions or pockets of hydrophobicity, or they
may be stabilized by combinations of these forces.

AMINO ACID SEQUENCES IN THERMOSTABLE PROTEINS

Ferredoxin

Ferredoxin has been purified from the thermophiles Clostridium tartarivorum and Clostridium thermosaccharolyticum,
and its properties have been compared to ferredoxin from the
mesophiles Clostridium pasteurianum, Clostridium acidiurici,
and Clostridium butyricum (Tanaka et al., 1971;

Devanathan *et al.*, 1969). On the basis of nearly all comparisons made, the ferredoxins from the thermophiles and the
mesophiles are identical. These include molecular weight,
iron, inorganic sulfide and cysteine content, absorption
spectra, and the number of amino acid residues. Despite
these many similarities, the thermophilic ferredoxins are
much more heat stable than the mesophilic ferredoxins. This
greater heat stability is dependent upon the structure of
apoferredoxin. Thus, iron and sulfide can be removed by extended heating, and the protein can then be reconstituted to
the original ferredoxin by adding back iron and sulfide.
This reconstituted ferredoxin has the same thermal stability
as the original (Devanathan *et al.*, 1969). As there is little
or no secondary structure in ferredoxin, the difference
between the mesophilic and thermophilic ferredoxins must
depend on the amino acid composition and, further, on the
sequence of amino acids.

The sequences of ferredoxin from C. tartarivorum and
three mesophilic clostridia are shown in Figure 1. The thermophilic clostridial ferredoxins are the only ferredoxins
that contain histidine (2 moles/mole of enzyme) (Devanathan
et al., 1969). It is possible that the histidines serve as
ligands for tighter binding of the iron atoms. Also, the
total number of basic amino acids in the thermophiles is
greater than in the mesophiles. In Figure 1, some of the
amino acids in the C. tartarivorum sequence are starred.
Tanaka *et al.* (1971) proposed that these amino acids are of
special importance to thermostability. Thus, His-2 replaces
Ser or Thr. Histidine has a different charge than the amino
acids it replaces, and may enter in H-bond formation. Glu-7
replaces Ala or Ser and Glu-25 replaces Gln. Glutamic acid,
of course, may stabilize the protein by ionizable group interaction. Other important changes are Thr-27 replacing aspartic
acid, glutamine-44 replacing Ala, Thr-49 replacing Val, Val-52
replacing Pro, and finally Lys-53 replacing Val or asparagine.

Thus, it becomes apparent that there are sequence differences between the thermophilic and mesophilic ferredoxins.
However, to directly relate these differences to higher thermostability, more information is needed on the precise three-
dimensional structure of the ferredoxins. The three-dimensional structure of the ferredoxin from P. aerogenes has been
determined from an electron density map at 2.8Å resolution
(Adman *et al.*, 1973). It is shown in Figure 2. The P. aerogenes ferredoxin differs from the thermostable ferredoxin
from C. tartarivorum, in that 22 amino acids have been

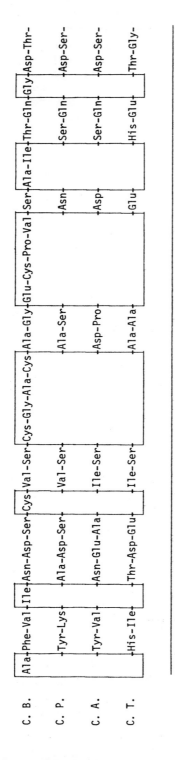
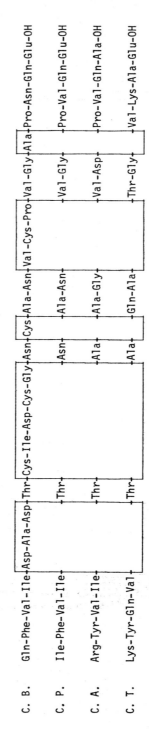

Fig. 1. A comparison of the primary structures of several ferredoxins. C.B., C.P., C.A., C.T., represent C. butyricum, C. pasteurianum, C. acidiurici, and C. tartarivorum ferredoxins, respectively. The numbers refer to residues starting from the amino-terminal residue. The constant residues are blocked off. [From Tanaka et al. (1971).]

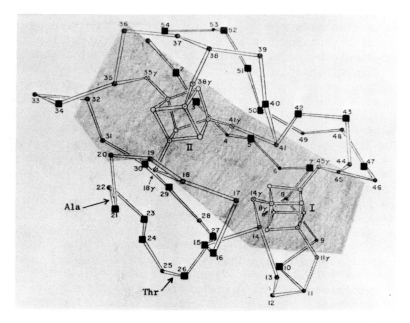

Fig. 2. *The three-dimensional structure of ferredoxin from* Peptococcus aerogenes, *as modified by us. Solid squares indicate amino acid substitution in the thermophilic ferredoxin, and the shaded area shows invariant regions of the ferredoxins.* [From Adman et al. (1973).]

substituted, and that the thermostable ferredoxin contains two additional amino acids (Ala between residues 21 and 22, and Thr between residues 25 and 26) and lacks the carboxy terminal residue (Asp-54). These changes are indicated in Figure 2 by heavy squares. It is interesting to see that the changes, almost without exception, have occurred outside the shaded area, which contains the active site of the ferredoxin. Residues 2, 3, and 5 may be behind and thus also outside the active site area. One can conclude that the amino acid residues responsible for the higher thermostability of the ferredoxin from C. tartarivorum must reside outside the active site area.

Cytochrome c_A

The amino acid sequence of cytochrome c prepared from several different sources is known. Figure 3 shows the sequence of cytochrome c_A from Humicola lanuginosa, a thermophilic fungus, which thrives at 60°C but does not grow below 30°C (Morgan *et al.*, 1972). The H. lanuginosa cytochrome c_A is compared with Neurospora crassa cytochrome *c*. They differ in 18 amino acids and in the methylation of lysine residues in two places, Lys-72 and Lys-86. The most remarkable changes are Ala-3 and Ala-55 in the thermophile, which are replaced with lysines. This is surprising, since it appears that alanine is generally present in higher amounts in mesophilic proteins, and lysine seems to occur more in thermophilic proteins. Perhaps the Lys-103 which replaces Ala is important for thermal stability. The methylated lysines are of great interest. The methylation changes the charge of the lysine amino group and may also contribute to hydrophobic bonds. Other noteworthy changes are Phe-74 for Tyr and Tyr-97 for Phe. Morgan and Riehm (1973) have done extensive studies of the H. lanuginosa cytochrome c_A using spectrophotometric titrations, optical rotatory dispersion (ORD), and circular dichroism (CD). They found that one or more of the Tyr residues 46, 48, 77 and 97, are buried in the native protein, and that Phe-74 is less exposed, due to its more hydrophobic character compared with the tyrosine residue it replaces. Although it is possible to suggest, knowing the amino acid sequence, that certain amino acids may contribute to a higher thermal stability, it is essential to know the three-dimensional structure to understand the interaction of the amino acid side chains in the H. lanuginosa cytochrome c_A.

Glyceraldehyde-3-phosphate Dehydrogenase

Glyceraldehyde-3-phosphate dehydrogenase isolated from B. stearothermophilus was the first intracellular enzyme crystallized from a thermophilic bacterium (Singleton and Amelunxen, 1973). The enzyme has been well-characterized physically and chemically (Singleton *et al.*, 1969; Amelunxen *et al.*, 1970; Bridgen *et al.*, 1972; Sauvan *et al.*, 1972). This enzyme has also been characterized from several mesophilic sources. Thus, a critical comparison between the enzyme from thermophilic and mesophilic sources can be made.

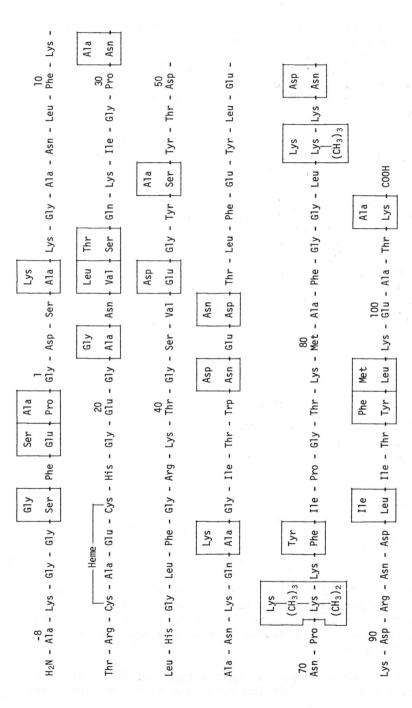

Fig. 3. *Amino acid sequence of* Humicola lanuginosa *cytochrome* c_A. *The residues in blocks, above those reported for the* H. lanuginosa *cytochrome* c_A *molecule, represent the residues for* Neurospora crassa *cytochrome* c. *[From Morgan et al. (1972).]*

161

The thermostable enzyme is active at 60°C and probably above, whereas the enzyme from mesophiles shows instability at 40°C. The enzyme from B. stearothermophilus, like the enzyme from rabbit muscle, lobster muscle, and pig muscle, is composed of four identical subunits, each with a molecular weight of 37,000, giving a total molecular weight of about 150,000. That there are four identical subunits, is supported by the observation that four moles of NAD$^+$ are covalently bound per mole of enzyme. The B. stearothermophilus enzyme contains four titratable —SH groups, while the muscle enzyme has 13-14 (Singleton et al., 1969). Of the 13-14 —SH groups in the rabbit muscle enzyme, it has been shown that there is one disulfide bond and one —SH group per each of four sub-units. The —SH group is believed to be at the active site of the enzyme. Thus, at the active site, both the muscle and B. stearothermophilus enzymes have a sulfhydryl group, and are similar, yet the overall number of —SH groups is differ-ent. For the muscle enzyme, it has been reported that NADH modifies the enzyme so that —SH groups are susceptible to oxidation by molecular oxygen (Amelunxen et al., 1970; Tucker and Grisolia, 1962). Similarly the B. stearothermophilus enzyme undergoes extensive inactivation in the presence of NADH, but only at elevated temperatures (Amelunxen, 1967). This substrate inactivation lends evidence to the similarity of the active sites.

Optical rotatory studies indicate structural differences between the muscle and B. stearothermophilus enzymes (Amelunxen et al., 1970). In 8.0 M urea, the latter enzyme does not undergo changes in optical rotation, and does not lose activity; by contrast the muscle enzyme shows an imme-diate loss of activity, and undergoes a large change in opti-cal rotation.

The amino acid composition of the B. stearothermophilus enzyme is quite similar to that of the enzyme from mesophilic sources. However, the thermostable enzyme has a decreased level of phenylalanine, which is offset by an increased level of leucine. Similarly, the level of methionine is decreased, whereas the level of alanine is increased. The combined total of aspartic acid and glutamic acid is higher in the thermostable enzyme than in the enzyme from muscles. Other-wise, the number of hydrophobic residues and proline are about the same.

It has been suggested, on the basis of ORD involving denaturant effects, that the catalytic site may contribute

significantly to the thermostability of the enzyme from
B. stearothermophilus (Sauvan *et al.*, 1972). The sequence of
a 17-residue peptide containing the catalytically active cysteine residue has been shown to be essentially identical in
all of the 14 mesophilic species that have been studied
(Perham, 1969). The tryptic peptide containing the active-
site cysteine has also been isolated from B. stearothermophilus (Bridgen *et al.*, 1972). The sequence of this peptide
is very similar to the comparable peptide from mesophilic
sources (Figure 4). However, at the *N*-terminus there is the

```
                                        1   2   3   4   5   6   7   8   9  10  11  12  13  14  15  16  17
Chicken, Ostrich, Sturgeon
Rabbit, Pig, Yeast, Ox              Ile-Val-Ser-Asn-Ala-Ser-Cys-Thr-Thr-Asn-Cys-Leu-Ala-Pro-Leu-Ala-Lys

Human                               Ile-Ile-Ser-Asn-Ala-Ser-Cys-Thr-Thr-Asn-Cys-Leu-Ala-Pro-Leu-Ala-Lys

Halibut                             Val-Val-Ser-Asn-Ala-Ser-Cys-Thr-Thr-Asn-Cys-Leu-Ala-Pro-Leu-Ala-Lys

Lobster                     Asp-Met-Thr-Val-Val-Ser-Asn-Ala-Ser-Cys-Thr-Thr-Asn-Cys-Leu-Ala-Pro-Val-Ala-Lys

B. stearothermophilus       Ala-His-His-Ile-Val-Ser-Asn-Ala-Ser-Cys-Thr-Thr-Asn-Cys-Leu-Ala-Pro-Phe-Ala-Lys
```

*Fig. 4. Amino acid sequence around the reactive cysteine
residue of glyceraldehyde-3-phosphate dehydrogenase,
from several sources. [From Perham (1969) and
Bridgen et al. (1972).]*

addition of Ala-His-His, and a phenylalanine has been substituted for a leucine. The presence of the two adjacent histidine residues in the active-site peptide is an interesting
variation present in the B. stearothermophilus enzyme. That
this variation contributes to the thermostability of the
enzyme, cannot be determined at this point. Bridgen and
Harris (1973) reported at the Ninth International Congress of
Biochemistry, the complete sequence of glyceraldehyde-

3-phosphate dehydrogenase from B. stearothermophilus, and found a 70% homology with the yeast enzyme. A Lys-183 residue, implicated in the catalytic site in the rabbit enzyme, is replaced by arginine in the thermostable enzyme.

Of interest, is that Olsen *et al.* (1974) have determined the three-dimensional structure of glyceraldehyde-3-phosphate dehydrogenase from lobster from a 3-Å resolution, electron density map. The B side of NAD^+ is close to the essential Cys-149. A His-176 is also close to the active site, and probably serves as a base catalyst, and the Lys-183 binds the pyrophosphate of NAD^+. This report indicates that soon, the three-dimensional structure of the thermophilic enzyme from B. stearothermophilus, and perhaps the enzyme from Thermus aquaticus, will be known. The enzyme from T. aquaticus has been purified to homogeneity. Physical studies and amino acid sequence analysis show that this enzyme is homologous with its counterparts from mesophilic sources. It also contains Cys-149 at the active site, like the enzyme from B. stearothermophilus. The T. aquaticus enzyme is very stable with respect to heat and treatment with SDS and 8 *M* urea (Hocking and Harris, 1973).

Ribosomal Proteins

Extensive studies of the ribosomes from Escherichia coli and B. stearothermophilus have been conducted. The ribosomes from the two bacteria are very similar, and yet they differ in thermal stability, the ribosomes of the thermophile being substantially more thermostable. Evidence indicates that the higher thermal stability resides in the proteins of the ribosome, thus, studies of the ribosomal proteins may lead to an understanding of thermophilicity.

The structure of the bacterial ribosome has been reviewed by Kurland (1971). The ribosome consists of two subunits, the 50S and the 30S subunits, which make the functional 70S ribosome. The 30S subunit consists of 16S RNA and 21 proteins labeled S1-S21. The 50S subunit contains two RNA molecules, 23S RNA and 5S RNA, and at least 28 proteins, or perhaps as many as 34, which are labeled L1-L34. The uniqueness of each ribosomal protein has been demonstrated using tryptic fingerprinting and immunological cross reactions. The ribosomes from B. stearothermophilus and E. coli are remarkably similar, and fractionations of the ribosomes from the two bacteria give almost identical results. Electron microscope pictures of the ribosomes from the two bacteria are very similar (Tecce and Toschi, 1960; Bassel and Campbell, 1969). The structure of the 30S subunit has been elucidated through reassembling experiments, and an assembly map of the E. coli 30S ribosome has been constructed (Figure 5) (Mizushima and Nomura, 1970; Held et al., 1974). As seen,

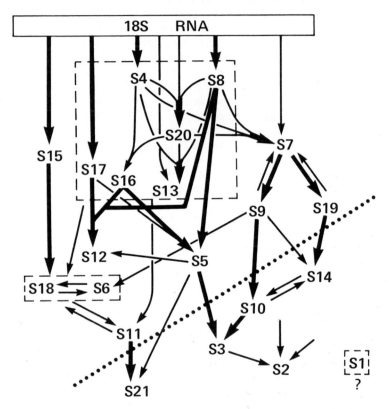

Fig. 5. An assembly map of E. coli 30S ribosomal proteins. [From Held et al. (1974).]

each of the 30S proteins has a special function and require- ment for binding. Many of the 30S ribosomal proteins from B. stearothermophilus were tested in the E. coli 30S ribosome reconstitution system and were found to replace corresponding E. coli proteins (Higo et al., 1973). Moreover, immunological cross reactions occurred between several E. coli and B.

stearothermophilus 30S ribosomal proteins, demonstrating not only functional, but also structural similarities between similar proteins from the 30S ribosomes of the two bacteria.

The thermostability of ribosomes from E. coli and B. stearothermophilus was first studied by Mangiantini et al. (1962). These studies have been extended by Saunders and Campbell (1966), Altenburg and Saunders (1971), and Pace and Campbell (1967). They showed that the melting temperatures of the 30S and the 50S subunits from B. stearothermophilus ribosomes are higher than those for the subunits from E. coli. More dramatic differences between E. coli and B. stearother- mophilus ribosomes were discovered by Friedmann (1968). He studied the ability of the ribosomes to catalyze poly-U- directed phenylalanine incorporation at 37°C, after the ribo- somes were exposed to 65°C. At 65°C the E. coli ribosomes rapidly lost their catalytic activity, while the activity of the B. stearothermophilus ribosomes was hardly affected.

Although the melting temperature of the 16S RNA from E. coli is somewhat lower than for the 16S RNA from B. stearo- thermophilus, the higher thermostability of the B. stearother- mophilus ribosome seems to reside, to a large degree, in the proteins. This is evident from the results obtained by Nomura et al. (1968). They reconstituted 30S ribosomes using RNA from E. coli and proteins from B. stearothermophilus, as well as RNA from B. stearothermophilus and proteins from E. coli. With RNA from E. coli or B. stearothermophilus, and proteins from E. coli, the reconstituted ribosomes rapidly lost activity when incubated at 65°C, and also at 55°C. How- ever, when the proteins were from B. stearothermophilus, there was almost no inactivation at 55°C, and activity was also observed after heating at 65°C. This occurred regard- less of the source of RNA. The results clearly demonstrated that the thermostability of the ribosomes depends mostly on the stability of the ribosomal proteins. A comparison between the homologous 30S ribosomal proteins from E. coli and B. stearothermophilus may, therefore, give clues regard- ing the basis of the higher thermostability of the B. stearo- thermophilus proteins.

The amino acid compositions of total 30S ribosomal pro- teins from B. stearothermophilus and E. coli have been deter- mined by Ansley et al. (1969), and they are remarkably simi- lar. Notable is the low content of cysteine (cystine), espe- cially in the protein from B. stearothermophilus. Ansley *et al.* (1969) also determined the amino acid composition of 12 fractions of 30S ribosomal proteins from B. stearothermophilus, but they were not paired with the equivalent proteins from E. coli. However, such comparison has been done for 10 of the ribosomal proteins from E. coli and B. stearothermophilus by Higo, Kahan and Vassos (personal communication). Although their results are preliminary, some observations may be related. First, the amino acid compositions between homologous proteins are very similar, and total differences of amino acid residues in molar percent, between the proteins, generally is in the order of 10. Many of the proteins lack cysteine (cystine), and this is especially evident for the proteins from B. stearothermophilus. Although the amino acid compositions are very similar, there are some noteworthy trends. Several of the proteins in B. stearothermophilus have more lysine and threonine, but less arginine residues than the E. coli proteins. Glycine is also more prominent in some of the B. stearothermophilus proteins, and appears to replace alanine in the E. coli proteins. The isoleucine and tyrosine contents are also higher in some of the thermostable proteins, which seems to be correlated with a lower content of valine and perhaps also cysteine residues, compared with the proteins from E. coli.

The amino acid sequences of the first 25-30 residues have been published for S4, S9, S10, S16, and S20 for the proteins from both E. coli and B. stearothermophilus by Higo and Loertscher (1974), and for the first 12 residues from S3, S9, and S13 by Yaguchi *et al.* (1973). These sequences are shown in Figure 6.

Although the complete sequences still remain to be elucidated, it is clear that there exists a great similarity between the functionally equivalent 30S proteins from E. coli and B. stearothermophilus. Many of the amino acid differences between the proteins can be explained by single base changes in their structural genes, and it is clear that the proteins have conserved regions which may function directly in their interactions in the ribosome. Perhaps the answer to the greater thermostability lies in the less conserved regions in which amino acids have been substituted. This may enhance the thermostability of the B. stearothermophilus proteins, compared with the equivalent proteins from E. coli. It should be mentioned that amino acid sequences are known for all except two of the 30S E. coli proteins, up to 34 and 60 residues (Wittman-Liebold, 1973), and the total sequence is known for protein S4 (Reinbolt and Schiltz, 1973) and the tryptic peptides of S12 (Funatsu *et al.*, 1972).

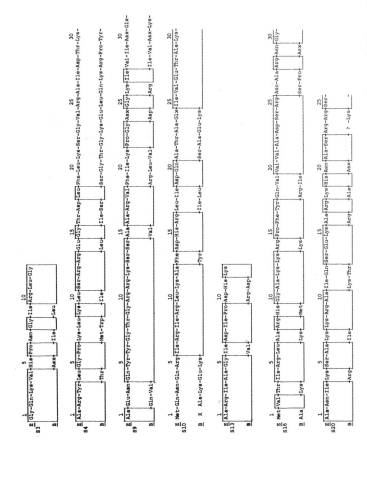

Fig. 6. *Comparison of partial amino acid sequences of seven pairs of 30S-ribosomal proteins from* E. coli *(E) and* B. stearothermophilus *(B). Homologous regions are boxed. Asx and Glx stand for aspartic acid or asparagine, and glutamic acid or glutamine, respectively. Note that the first residue of B S10 aligns with position 2 of E S10. [From Higo et al. (1974) and Yaguchi et al. (1973).]*

168

Reconstitution of the 50S ribosomal subunits has been achieved by Nomura and coworkers (Fahnestock *et al.*, 1973; Nomura and Erdmann, 1970) for the B. stearothermophilus subunit, but not for the E. coli subunit. A possible explanation for the lack of results with the E. coli subunit may lie in its lower temperature stability. The activation energy for the reconstitution of the E. coli 30S subunit is about 40 cal/mole, and it is possible that the reconstitution of the 50S subunit has a much higher free energy of activation. Therefore, a higher temperature may be required for the reconstitution of the 50S subunit, and this higher temperature may inactivate the 50S components from E. coli. The reconstitution of the B. stearothermophilus 50S subunit was done at 60°C.

Irwin *et al.* (1973) have investigated ribosomes, polyribosomes and DNA from anaerobic bacteria. Two are thermophilic, C. thermosaccharolyticum and C. tartarivorum; one is mesophilic, C. pasteurianum W 5; and one is psychrophilic, Clostridium sp. strain 69. The T_m of ribosomes for the two thermophiles are the same (68.7°C and 69.6°C), while that of the mesophile is 63.8°C, and of the psychrophile 64.4°C. The T_m of the rRNA from all four clostridia are very similar (66.2°-67.9°C). There are some qualitative differences in the "melting out" curves, in that some melting out occurs at lower temperatures, with the ribosomes from C. pasteurianum and Clostridium sp. strain 69.

The functionality of the ribosomes from the four clostridia was tested at 37°C after incubations for various times at 55°C, and by preincubations for five min at different temperatures. At 55°C ribosomes from the psychrophilic organism lost all activity within 10 min. C. pasteurianum ribosomes lost activity gradually, and 8% of the original activity remained after 1 hour at 55°C. The ribosomes from C. thermosaccharolyticum lost only 25% of the activity by incubation for 1 hour at 55°C, while the ribosomes from C. tartarivorum were fully active after 1 hour at 55°C. The temperature at which 50% of the ribosomal activity remained after preincubation for 5 min (T_d.5) is for Clostridium sp. strain 69, 53°C; C. pasteurianum, 57°C; C. tartarivorum, 70°C, and C. thermosaccharolyticum, 72°C.

Optical rotatory dispersion spectra of the ribosomes from C. thermosaccharolyticum, C. pasteurianum, and Clostridium sp. strain 69 are nearly superimposable and comparable with spectra from E. coli ribosomes. The amino acid compositions of the ribosomes from the clostridia are almost identical to that of ribosomes from E. coli.

Based on these and other data Irwin *et al.* (1973) concluded that ribosomes from thermophilic clostridia are more heat stable than ribosomes from clostridia growing at lower temperatures. Furthermore, they suggested that the rRNA does not contribute to the greater thermal stability of the ribosomes from the thermophiles; instead, they suggested that the ribosomal proteins may be the important factor for determining the thermal stability of the ribosomes. Work similar to that discussed above, regarding the ribosomal proteins from B. stearothermophilus, should perhaps clarify this point.

Other parts of the protein-synthesizing machinery in thermophiles have been purified. Thus the tyrosyl-tRNA synthetase (Koch, 1974) and the valyl-tRNA synthetase (Wilkinson and Knowles, 1974) have both been purified from B. stearothermophilus, and their properties have been compared with their counterparts from E. coli. The similarities are striking, but the B. stearothermophilus proteins are more thermostable. A DNA-dependent RNA-polymerase has also been purified from B. stearothermophilus, and, again, this enzyme has a considerably higher heat stability as compared with the E. coli enzyme (Remold-O'Donnell and Zillig, 1969).

Rates of Reactions Catalyzed by Enzymes from Thermophiles

The rates of enzyme reactions normally double by performing them at a 10-degree higher temperature. If an enzyme from a thermophile had the same catalytic activity as an enzyme from a mesophile, one would expect the enzyme from the thermophile to have an extremely high activity at the temperature at which the thermophile grows. For instance, the pure formyltetrahydrofolate synthetase from the mesophile C. cylindrosporum has an activity of 396 μmoles/min/mg at 37°C (Rabinowitz and Pricer, 1962). If the enzyme from C. thermoaceticum had the same activity as the enzyme from the mesophile, one would expect that at 57°C, it would catalyze the reaction at a rate of 1600 μmoles/min/mg. This is not what one observes. The C. thermoaceticum enzyme has an activity of only 230 μmoles/min/mg at 50°C (Ljungdahl *et al.*, 1970). Similarly, most enzymes from thermophiles do not have exceptionally high activities, compared with their counterparts from mesophiles. It thus appears that the enzymes from thermophiles have sacrificed part of their catalytic activities for higher thermal stability. Thus, it follows that many enzymes from thermophilic organisms have very low activities at temperatures below 40°C. This may explain why many thermophiles stop growing at temperatures below 40°C.

Many enzymes appear to undergo conformational changes going from low to high temperatures, as evidenced by broken Arrhenius plots or plots of K_m against temperature. For a discussion of nonlinear Arrhenius plots, see Han (1972). Broken Arrhenius plots of K_m against temperature is not a phenomenon exclusive to enzymes from thermophilic organisms. Many proteins from mesophilic organisms also show broken plots. Examples of such enzymes are found in Table 3. With most enzymes, the activation energy calculated from the Arrhenius plots is high at lower temperatures and low at high temperatures. Thus, the enzyme reaction must overcome a higher activation energy at low temperatures. This will undoubtedly slow down the metabolism, and may partially explain why many thermophilic organisms do not grow, or grow very slowly at low temperatures. There are, however, examples of enzymes having a higher activation energy at high, rather than at low temperatures. Such an example is the ribonucleotide reductase from Thermus X-1. It has an activation energy below 45°C of 13.3 cal/mole and above 45°C of 17.9 cal/mole (Sando and Hogenkamp, 1973). This enzyme is highly thermostable.

Formyltetrahydrofolate Synthetase: Subunit Interaction; a Possible Mechanism of Thermal Stability

Formyltetrahydrofolate synthetase, from C. cylindrosporum and C. acidiurici, has been studied extensively by Himes and Rabinowitz, and their coworkers. The enzymes from C. formicoaceticum and C. thermoaceticum have also been studied. A review of the properties of formyltetrahydrofolate synthetase was recently published by Himes and Harmony (1973). The thermostable enzyme from C. thermoaceticum is remarkably similar to the enzyme from the three mesophilic clostridia (Ljungdahl et al., 1970; Brewer et al., 1970). The enzyme is a tetramer having a molecular weight of 240,000. The enzyme from the mesophiles requires potassium or ammonium ions for activity. Removal of these ions causes a dissociation of the enzyme into subunits (MacKenzie and Rabinowitz, 1971; Welch et al., 1971). Reconstitution is accomplished by adding potassium or ammonium ions back to the solution of the subunits. The enzyme from the thermophile is not only more thermostable, but it also maintains its tetrameric

Table 3

Enzymes Exhibiting Broken Arrhenius or K_m versus Temperature Plots

Enzyme	Broken Arrhenius	Broken K_m plot	Transition temperature (°C)	Reference
Lipase	+		0°	Sizer and Josephson (1942).
D-amino acid oxidase (pig kidney)	+	+	12°-14°	Massey et al. (1966).
Fumarase (pig heart)	+	+	12°-27°	Massey (1953).
Aldolase (rabbit muscle)	+	+	25°	Lehrer and Barker (1970).
Formyl tetrahydrofolate synthetase (C. cylindrosporum)		+	30°	Himes and Wilder (1968).
Uridine kinase (Novikoff tumor cells)	+		32°-34°	Orengo and Saunders (1972).
DNA-dependent RNA polymerase (B. stearothermophilus)	+		33°	Remold-O'Donnell and Zillig (1969).
Methylene tetrahydrofolate dehydrogenase (C. thermoaceticum)	+		35°	O'Brien et al. (1973).

Table 3 (continued)

Enzymes Exhibiting Broken Arrhenius or K_m versus Temperature Plots

Enzyme	Broken Arrhenius plot	Broken K_m plot	Transition temperature (°C)	Reference
Formyl-tetrahydrofolate synthetase (C. thermoaceticum)	+	+	40°–43°	Shoaf et al. (1974).
Uridine kinase (B. stearothermophilus)	+		45°–47°	Orengo and Saunders (1972).
Aldolase (B. stearothermophilus)	+	+	50°–53°	Hachimori et al. (1970).
ATPase (B. stearothermophilus)	+	+	55°	Weerkamp and MacElroy (1972).
Glucose-6-phosphate isomerase (B. stearothermophilus)		+	55°	Muramatsu and Nosoh (1971).

structure after removal of the monovalent cations. However, ammonium or potassium ions do affect the thermostable enzyme, probably by causing a conformational change (Shoaf *et al.*, 1974). The C. cylindrosporum has a broken K_m against temperature plot (Himes and Harmony, 1973), with the break at about 30°C. The C. thermoaceticum enzyme exhibits a similar break in the K_m against temperature plot, but the break is at 45°C. In addition, the K_m values of all three substrates are affected by ammonium ions, as shown in Figure 7. The Arrhenius plot has a break also, at about 45°C (Figure 8), and the slopes of the curves are changed when ammonium ions are added. Although the kinetic data indicate a conformational

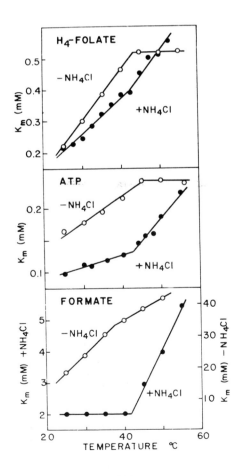

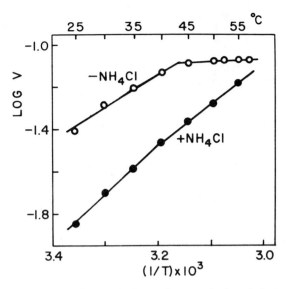

Fig. 8. *Log of velocity of formyltetrahydrofolate synthetase*
as a function of the reciprocal of the absolute tem-
perature. *[From Shoaf et al. (1974).]*

change, ORD studies did not reveal such a change, thus indicating that the change should be very small. A small conformational change in the C. thermoaceticum enzyme is supported by a break in the absorbance at 295 nm versus temperature (Shoaf *et al.*, 1974). The break occurs at about 45°C,
and again is affected by ammonium ions. The fluorescence and
the reactivity of the enzyme with *p*-hydroxymercuribenzoate
also are affected by temperature and ammonium ions. These
results have led us to postulate that the C. thermoaceticum
enzyme exists in several conformational forms, depending on

Fig. 7. *Apparent K_m values of the substrates in the forward*
reaction as a function of temperature, and the ef-
fect of NH_4Cl. *[From Shoaf et al. (1974).]*

temperature and on the presence or absence of ammonium ions (Shoaf *et al.*, 1974).

It is of interest that Buttlaire *et al.* (1972) have been able to dissociate the C. cylindrosporum enzyme at pH 5.3 into unfolded subunits, which reassociate. The C. thermoaceticum enzyme is also dissociated in the vicinity of pH 5. However, the subunits are no longer able to reassociate. This may indicate that the subunits from C. thermoaceticum are less stable than those of the C. cylindrosporum enzyme. If this is correct, the higher temperature stability of the C. thermoaceticum enzyme may depend on the association of the subunits which stabilize each other. The interaction between the subunits may be hydrophobic. This is supported by data presented by Welch *et al.* (1971) who found with the C. cylindrosporum enzyme, that tryptophan, tyrosine and phenylalanine residues become buried in hydrophobic regions, when subunits reassociate.

Koffler and Yarbrough (1972) have reported about thermostable flagellar filaments from B. stearothermophilus. They found that disintegration of the filaments does not begin until temperatures are reached, at which flagellins are largely unfolded, and that the unfolding is responsible for the disintegration of the polymer. The flagellin may be modified by reacting tyrosine residues with tetranitromethane. The resulting flagellin will reassemble to form filaments. These filaments disintegrate at lower temperatures and yield folded monomers. It was suggested that the hydrophobic bonds are participating in the flagellin-flagellin interaction, and that the tetranitromethane treatment modified the hydrophobic interaction.

That tetrameric structures like the formyltetrahydrofolate synthetase from C. thermoaceticum appear to be more stable than the subunits, suggests that peptide chain interactions stabilize proteins. Similarly, Wedler (1972) suggests that subunit aggregation is the basis of thermostability in glutamine synthetase from B. stearothermophilus. It is also possible that association of different proteins may have higher thermostability than corresponding single proteins. Donovan and Beardslee (1974) have presented evidence that association complexes of different proteins have higher thermostability. It is also to be noted that enzymes fixed to a stationary phase, such as Agarose or Sephadex, often have higher thermal stability when compared with the free enzyme (Porath, 1972). If associations between different proteins

are more heat stable, this may explain why some purified enzymes from obligate thermophilic and facultative thermophilic
organisms have a thermal stability less than the growth temperature of the organism.

ALLOSTERISM AND FEEDBACK CONTROL OF ENZYMES FROM THERMOPHILES

Brock (1967) has suggested that thermostable enzymes
have rigid, inflexible structures, and that such enzymes have
sacrificed efficiency and control of enzyme function, for
thermostability. Thus, thermostable enzymes should not be
able to undergo allosteric conformational changes. However,
during the last few years, several enzymes have been
isolated from thermophilic organisms that have allosteric,
or feedback controls. The enzymes can be divided into two
groups: those enzymes which have allosteric effects
both at low and high temperatures, and those enzymes which
have allosteric effects only at high temperatures. Table 4
shows enzymes that belong to the first group. Aspartokinase
has been isolated from both B. stearothermophilus (Cavari et
al., 1972; Kuramitsu, 1968; Kuramitsu, 1970) and from a
bacterium similar to T. aquaticus or Flavobacterium thermophilum (Saiki and Arima, 1970). Phosphofructokinase (Yoshida
et al., 1971; Yoshida, 1972) and fructose 1,6-diphosphatase
(Yoshida and Oshima, 1971) are from F. thermophilum. Threonine deaminase is from B. stearothermophilus.

Table 5 shows examples of the second group of allosteric
thermostable enzymes. They show allosteric effects only at
high temperatures, temperatures at which the bacteria grow
naturally. Lactate dehydrogenase is from Bacillus caldolyticus (Massey, 1953). Pyrimidine ribonucleoside kinase
(Orengo and Saunders, 1972) has been isolated from B. stearothermophilus, and homoserine dehydrogenase (Cavari and
Grossowicz, 1973) from a similar bacterium, although it is
unable to hydrolyze starch. The B_{12}-dependent ribonucleotide
reductase is from Thermus X-1 (Sando and Hogenkamp, 1973),
and acetohydroxy acid synthetase comes from T. aquaticus and
a Bacillus sp. (Chin and Trela, 1973).

The existence of thermostable proteins which show allosteric effects only at high temperatures, may indicate, as
suggested by Brock, that these enzymes are more rigid than
their counterparts from mesophilic origin. However, they
are rigid only at low temperatures. At the higher temperatures at which the thermophilic bacteria grow, the enzymes
are subject to allosteric control. Apparently, there is

177

Table 4

Enzymes Having Allosteric or Feedback Control

Enzyme	Organism	Special feature	Reference
Aspartokinase	B. stearothermophilus	Lysine, Threonine inh.	Kuramitsu (1968).
	Thermus sp.	Synergistic effect	Saiki and Arima (1970).
Threonine deaminase	B. stearothermophilus	Isoleucine inh.	Thomas and Kuramitsu (1971).
Phosphofructokinase	F. thermophilum	PEP inh.	Yoshida et al. (1971); Yoshida (1972).
Fructose 1,6-diphosphatase	F. thermophilum	AMP inh. PEP act.	Yoshida and Oshima (1971).
Homoserine dehydrogenase	B. stearothermophilus sp.	Threonine inh.	Cavari and Grossowicz (1973).

Table 5

Enzymes with Allosteric or Feedback Control at High Temperatures, But Not at Low

Enzyme	Organism	Special feature	Reference
Lactate dehydrogenase	B. caldolyticus	ADP, PEP inh.	Weerkamp and MacElroy (1972).
Uridine kinase	B. stearothermophilus	CTP inh.	Orengo and Saunders (1972).
Ribonucleotide reductase	Thermus X-1	dGTP act. ATP reduction	Sando and Hogenkamp (1973).
Acetohydroxy-acid synthetase	T. aquaticus Bacillus sp.	Valine, Isoleucine inh.	Chin and Trela (1973).

enough conformational entropy change, due to the higher temperature, to make the proteins flexible enough for allosteric control.

CONCLUSION

Data presented indicates that proteins from thermophilic organisms behave in all aspects like similar proteins from mesophilic organisms, except that they are generally more thermostable. However, there are differences in amino acid content, and in the sequence. The variation in thermostability must depend on these differences, although they are subtle. Wilkinson and Knowles (1974) suggest that searches for characteristic differences in amino acid composition between proteins from mesophiles and thermophiles, may well be fruitless, especially as differences in stability are very small, energetically and structurally. However, it must be considered that with our greater capacity to sequence proteins and to look into their three-dimensional structures, explanations will certainly be found for the differences in thermostability between homologous proteins. Furthermore, studies of thermostable proteins will lead to a better understanding of proteins, in general, and probably will show the importance of the part of the amino acid sequence outside the active site.

Recently Chou and Fasman (1974a,b) showed that conformational parameters, such as helical β-sheet, and random coil regions can be calculated from the amino acid composition. Such calculations may also help us to understand how certain amino acids may affect the thermostability of proteins.

Although it is now well established that proteins from thermophiles are more thermostable, compared to their counterparts from mesophiles, this does not mean that we know, in broad terms, the basis for thermophilicity. Certainly, other components of the thermophilic cells also need modifications, to function at the high temperatures at which the thermophiles grow. Furthermore, the fundamental question as to how thermophilic microorganisms evolved, still remains to be answered.

ACKNOWLEDGMENT

We wish to express our appreciation to Dr. L. Kahan for communicating the amino acid compositions of ribosomal proteins, before publication. Work presented from our laboratory was supported by the Public Health Service Grant AM-12913 from the National Institute of Arthritis and Metabolic Diseases, and by Grant GB-13031 from the National Science Foundation.

REFERENCES

ADMAN, E. T., SIEKER, L. C., and JENSEN, L. H. (1973). J. Biol. Chem. 248, 3987.

ALLEN, M. B. (1953). Bacteriol. Rev. 17, 125.

ALTENBURG, L. C., and SAUNDERS, G. F. (1971). J. Mol. Biol. 55, 487.

AMELUNXEN, R. E. (1967). Biochim. Biophys. Acta 139, 24.

AMELUNXEN, R. E., and LINS, M. (1968). Arch. Biochem. Biophys. 125, 765.

AMELUNXEN, R. E., NOELKEN, M., and SINGLETON, Jr., R. (1970). Arch. Biochem. Biophys. 141, 447.

ANDREESEN, J. R., SCHAUPP, A., NEURAUTER, C., BROWN, A., and LJUNGDAHL, L. G. (1973). J. Bacteriol. 114, 743.

ANSLEY, S. B., CAMPBELL, L. L., and SYPHERD, P. S. (1969). J. Bacteriol. 98, 568.

BARNES, L. D., and STELLWAGEN, E. (1973). Biochemistry 12, 1559.

BASSEL, A., and CAMPBELL, L. L. (1969). J. Bacteriol. 98, 811.

BIGELOW, C. C. (1967). J. Theoret. Biol. 16, 187.

BRANDTS, J. F. (1967). In "Thermobiology" (A. H. Rose, ed.), pp. 25-72. Academic Press, New York.

BREWER, J. M., LJUNGDAHL, L., SPENCER, T. E., and NEECE, S. H. (1970). J. Biol. Chem. 245, 4798.

BRIDGEN, J., and HARRIS, J. I. (1973). 9th Intern. Congr. Biochem., Stockholm, Commun. 2el.

BRIDGEN, J., HARRIS, J. I., McDONALD, P. W., AMELUNXEN, R. E., and KIMMEL, J. R. (1972). J. Bacteriol. 111, 797.

BROAD, T. E., and SHEPHERD, M. G. (1970). Biochim. Biophys. Acta 198, 407.

BROCK, T. D. (1967). Science 158, 1012.

BROCK, T. D., and FREEZE, H. (1969). J. Bacteriol. 98, 289.

BUTTLAIRE, D. H., HERSH, R. T., and HIMES, R. H. (1972). J. Biol. Chem. 247, 2059.

CAMPBELL, L. L., and MANNING, G. B. (1961). J. Biol. Chem. 236, 2962.

CASTENHOLZ, R. W. (1969). Bacteriol. Rev. 33, 476.

CAVARI, B. Z., and GROSSOWICZ, N. (1973). Biochim. Biophys. Acta 302, 183.

CAVARI, B. Z., ARKIN-SHLANK, H., and GROSSOWICZ, N. (1972). Biochim. Biophys. Acta 261, 161.

CHIN, N. W., and TRELA, J. M. (1973). J. Bacteriol. 114, 674.

CHOU, P. Y., and FASMAN, G. D. (1974a). Biochemistry 13, 211.

CHOU, P. Y., and FASMAN, G. D. (1974b). Biochemistry 13, 222.

COONEY, D. G., and EMERSON, R. (1964). "Thermophilic Fungi." Freeman, San Francisco.

DEVANATHAN, T., AKAGI, J. M., HERSH, R. T., and HIMES, R. H. (1969). J. Biol. Chem. 244, 2846.

DIXON, M., and WEBB, E. C. (1964). In "Enzymes," pp. 145-166. Academic Press, New York.

DONOVAN, J. W., and BEARDSLEE, R. A. (1974). Fed. Proc. 33, 1504, Abstr. 1581.

FAHNESTOCK, S., ERDMANN, V., and NOMURA, M. (1973). Biochemistry 12, 220.

FARRELL, J., and CAMPBELL, L. L. (1969). Advan. Microbiol. Physiol. 3, 83.

FEDER, J., GARRETT, L. R., and WILDI, B. S. (1971). Biochemistry 10, 4552.

FISHER, H. F. (1964). Proc. Nat. Acad. Sci. U.S.A. 51, 1285.

FREEZE, H., and BROCK, T. D. (1970). J. Bacteriol. 101, 541.

FRIEDMANN, S. M. (1968). Bacteriol. Rev. 32, 27.

FUNATSU, G., NIERHAUS, K., and WITTMAN, H. G. (1972). Biochim. Biophys. Acta 287, 282.

GOLDSACK, D. E. (1970). Biopolymers 9, 247.

HACHIMORI, A., MURAMATSU, N., and NOSOH, Y. (1970). Biochim. Biophys. Acta 206, 426.

HAN, M. H. (1972). J. Theor. Biol. 35, 543.

HELD, W. A., BALLOU, B., MIZUSHIMA, M., and NOMURA, M. (1974). J. Biol. Chem. 249, 3103.

HIGO, K.-I., and LOERTSCHER, K. (1974). J. Bacteriol. 118, 180.

HIGO, K.-I., HELD, W., KAHAN, L., and NOMURA, M. (1973). Proc. Nat. Acad. Sci. U.S.A. 70, 944.

HIGO, K.-I., KAHAN, L., VASSOS, A. (personal communication from L. Kahan).

HIMES, R. H., and HARMONY, J. A. K. (1973). CRC, Critical Reviews in Biochemistry, Sept. 501.

HIMES, R. H., and WILDER, T. (1968). Arch. Biochem. Biophys. 124, 230.

HOCKING, J. D., and HARRIS, J. I. (1973). 9th Intern. Congr. Biochem., Stockholm, Commun. 2c8.

HOWARD, R. L., and BECKER, R. R. (1970). J. Biol. Chem. 245, 3186.

HOWARD, R. L., and BECKER, R. R. (1972). Biochim. Biophys. Acta 268, 249.

HOWELL, N., AKAGI, J. M., and HIMES, R. H. (1969). Canad. J. Microbiol. 15, 461.

IRWIN, C. C., AKAGI, J. M., and HIMES, R. H. (1973). J. Bacteriol. 113, 252.

ISONO, K. (1970). Biochem. Biophys. Res. Commun. 41, 852.

KOCH, G. L. E. (1974). Biochemistry 13, 2307.

KOFFLER, H., and YARBROUGH, L. R. (1972). Conf. on Extreme Environments. Ames Research Center, NASA, Moffett Field. CA.

KURAMITSU, H. K. (1968). Biochim. Biophys. Acta 167, 643.

KURAMITSU, H. K. (1970). J. Biol. Chem. 245, 2991.

KURLAND, C. G. (1971). Ann. Rev. Biochem. 41, 377.

LEHRER, G. M., and BARKER, R. (1970). Biochemistry 9, 1533.

LJUNGDAHL, L., IRION, E., and WOOD, H. G. (1965). Biochemistry 4, 2771.

LJUNGDAHL, L., BREWER, J. M., NEECE, S. H., and FAIRWELL, T. (1970). J. Biol. Chem. 245, 4791.

MacKENZIE, R. E., and RABINOWITZ, J. C. (1971). J. Biol. Chem. 246, 3731.

MANGIANTINI, M. T., TEECE, G., and TOSCHI, G. (1962). Nuovo Cimento 25, Suppl. 45.

MANNING, G. B., CAMPBELL, L. L., and FOSTER, R. J. (1961). J. Biol. Chem. 236, 2958.

MASSEY, V. (1953). Biochem. J. 53, 72.

MASSEY, V., CURTI, B., and GANTHER, H. (1966). J. Biol. Chem. 241, 2347.

MILITZER, W., SONDEREGGER, T. B., TUTTLE, G. C., and GEORGI, C. E. (1949). Arch. Biochem. Biophys. 24, 75.

MIZUSAWA, K., and YOSHIDA, F. (1972). J. Biol. Chem. 247, 6978.

MIZUSAWA, K., and YOSHIDA, F. (1973). J. Biol. Chem. 248, 4417.

MIZUSHIMA, S., and NOMURA, M. (1970). Nature 226, 1214.

MOORE, M. R., O'BRIEN, W. E., and LJUNGDAHL, L. G. (1974). J. Biol. Chem. 249, 5250.

MORGAN, W. T., and RIEHM, J. P. (1973). Arch. Biochem. Biophys. 154, 415.

MORGAN, W. T., HENSLEY, Jr., C. P., and RIEHM, J. P. (1972). J. Biol. Chem. 247, 6555.

MOSER, R., RONCARI, G., and ZUBER, H. (1970). Int. J. Protein Res. 2, 191.

MURAMATSU, N., and NOSOH, Y. (1971). Arch. Biochem. Biophys. 144, 245.

NAKAMURA, Y. (1960). J. Biochem. (Tokyo) 48, 295.

NOMURA, M., and ERDMANN, V. A. (1970). Nature 228, 744.

NOMURA, M., TRAUB, P., and BECHMANN, H. (1968). Nature 219, 793.

O'BRIEN, W. E., and LJUNGDAHL, L. G. (1972). J. Bacteriol. 109, 626.

O'BRIEN, W. E., BREWER, J. M., and LJUNGDAHL, L. G. (1973). J. Biol. Chem. 248, 403.

OGASAHARA, K., IMANISHI, A., and ISEMURA, T. (1970) J. Biochem. (Tokyo) 67, 65.

OHTA, Y. (1967). J. Biol. Chem. 242, 509.

OHTA, Y., OGURA, Y., and WADA, A. (1966). J. Biol. Chem. 241, 5919.

OLSEN, K. W., BUEHNER, M., FORD, G. C., MORAS, D., and ROSSMAN, M. G. (1974). Fed. Proc. 33, 1374. Abstr. 852.

ORENGO, A., and SAUNDERS, G. F. (1972). Biochemistry 11, 1761.

PACE, B., and CAMPBELL, L. L. (1967). Proc. Nat. Acad. Sci. U.S.A. 57, 1110.

PERHAM, R. N. (1969). Biochem. J. 111, 17.

PFUELLER, S. L., and ELLIOT, W. H. (1969). J. Biol. Chem. 244, 48.

PORATH, J. (1972). Kemisk Tidskrift 84, Number 1-2, page 40.

POSTON, J. M., KURATOMI, K., and STADTMAN, E. R. (1966). J. Biol. Chem. 241, 4209.

RABINOWITZ, J. C., and PRICER, Jr., W. E. (1962). J. Biol. Chem. 237, 2898.

REINBOLT, J., and SCHILTZ, E. (1973). FEBS Letters 36, 250.

REMOLD-O'DONNELL, E., and ZILLIG, W. (1969). European J. Biochem. 7, 318.

185

RONCARI, G., and ZUBER, H. (1969). Int. J. Protein Res. 1, 45.

SAIKI, T., and ARIMA, K. (1970). Agr. Biol. Chem. 34, 1762.

SANDO, G. N., and HOGENKAMP, H. P. C. (1973). Biochemistry 12, 3316.

SAUNDERS, G. F., and CAMPBELL, L. L. (1966). J. Bacteriol. 91, 332.

SAUNDERS, P. P., WILSON, B. A., and SAUNDERS, G. F. (1969). J. Biol. Chem. 244, 3691.

SAUVAN, R. L., MIRA, O. J., and AMELUNXEN, R. E. (1972). Biochim. Biophys. Acta 263, 794.

SCHULMAN, M., GHAMBEER, R. K., LJUNGDAHL, L. G., and WOOD, H. G. (1973). J. Biol. Chem. 248, 6255.

SHOAF, W. T., NEECE, S. H., and LJUNGDAHL, L. G. (1974). Biochim. Biophys. Acta 334, 448.

SINGLETON, Jr., R., and AMELUNXEN, R. E. (1973). Bacteriol. Rev. 37, 320.

SINGLETON, Jr., R., KIMMEL, J. R., and AMELUNXEN, R. E. (1969). J. Biol. Chem. 244, 1623.

SIZER, I. W., and JOSEPHSON, E. S. (1942). Food Res. 7, 201.

SMYTH, D. G., STEIN, W. H., and MOORE, S. (1963). J. Biol. Chem. 238, 227.

SPACKMAN, D. H., STEIN, W. H., and MOORE, S. (1960). J. Biol. Chem. 235, 648.

STANIER, R. Y., DOUDOROFF, M., and ADELBERG, E. A. (1970). "The Microbial World," 3rd ed., pp. 315-318. Prentice-Hall, Inc., Englewood Cliffs, N. J.

STELLWAGEN, E., CRONLUND, M. M., and BARNES, L. D. (1973). Biochemistry 12, 1552.

STENESH, J., and ROE, B. A. (1972). Biochim. Biophys. Acta 272, 156.

STOLL, E., HERMODSON, M. A., ERICSSON, L. H., and ZUBER, H. (1972). Biochemistry 11, 4731.

186

SUGIMOTO, S., and NOSOH, Y. (1971). Biochim. Biophys. Acta 235, 210.

TANAKA, M., NAKASHIMA, T., BENSON, A., MOWER, H., and YASUNOBU, K. T. (1966). Biochemistry 5, 1666.

TANAKA, M., HANIU, M., MATSUEDA, G., YASUNOBU, K., HIMES, R. H., AKAGI, J. M., BARNES, E. M., and DEVANATHAN, T. (1971). J. Biol. Chem. 246, 3953.

TANFORD, C. (1962). J. Amer. Chem. Soc. 84, 4240.

TEECE, G., and TOSCHI, G. (1960). Nuovo Cimento 18, Suppl. 207.

THOMAS, D. A., and KURAMITSU, H. K. (1971). Arch. Biochem. Biophys. 145, 96.

THOMPSON, P. J., and THOMPSON, T. L. (1962). J. Bacteriol. 84, 694.

TUCKER, D., and GRISOLIA, S. (1962). J. Biol. Chem. 237, 1068.

WAUGH, D. F. (1954). Adv. in Protein Chemistry 9, 325.

WEDLER, F. C. (1972). Conf. on Extreme Environments. Ames Research Center, NASA, Moffett Field, CA.

WEERKAMP, A., and MacELROY, R. D. (1972). Arch. Mikrobiol. 85, 113.

WELCH, W. H., BUTTLAIRE, D. H., HERSH, R. T., and HIMES, R. H. (1971). Biochim. Biophys. Acta 236, 599.

WILKINSON, S., and KNOWLES, J. R. (1974). Biochem. J. 139, 391.

WITTMAN-LIEBOLD, B. (1973). FEBS Letters 36, 247.

YAGUCHI, M., ROY, C., MATHESON, A. T., and VISENTIN, L. P. (1973). Canad. J. Biochem. 51, 1215.

YOSHIDA, M. (1972). Biochemistry 11, 1087.

YOSHIDA, M., and OSHIMA, T. (1971). Biochem. Biophys. Res. Commun. 45, 495.

YOSHIDA, M., OSHIMA, T., and IMAHORI, K. (1971). Biochem. Biophys. Res. Commun. 43, 36.

A COMPARISON OF THE AMINO ACID COMPOSITIONS OF PROTEINS FROM THERMOPHILIC AND NON-THERMOPHILIC ORIGINS

Rivers Singleton, Jr.

As Dr. Ljungdahl's article has demonstrated, the major significant difference between proteins from thermophilic bacteria and those of mesophilic bacteria is their thermostability. Proteins from these organisms are *almost* unique in their capacity to withstand denaturation by heat. However, it now seems clear that the majority of these proteins have physical and chemical properties very much in common with their nonthermophilic counterparts. In addition, at the gross level, it appears that the amino acid compositions of thermophilic proteins are similar to those of the corresponding nonthermophilic proteins. However, most of these comparisons have been made with a small number of proteins, and therefore subtle differences in composition may have been overlooked.

It was the intent of the present study to obviate this problem by comparing the amino acid compositions of all of the thermophilic proteins characterized thus far. It was anticipated that if consistent subtle differences did occur in the amino acid composition of proteins from these organisms, they might be detected by this means.

To this end, data have been compiled of the amino acid

189

compositions of thermophilic proteins (Table 1). Included are the data for three α-amylases and two enolases. Table 1 also shows the number of nonthermophilic proteins used for comparison, broken down according to prokaryotic or eukaryotic origin. This breakdown is shown because Reeck and Fisher (1973) have demonstrated that a slight difference exists in the amino acid compositions for these two classes of organisms. However, they could not attribute this bias to any single amino acid. We felt it worth the risk of introducing this possible bias into our data pool to give as wide a data base as possible, for comparison.

The amino acid compositions were compiled from the

Table 1

Proteins from Thermophilic Bacteria
for Which Amino Acid Compositions are Known[a]

Thermophilic Enzyme	Number of Mesophilic Comparisons		
	Prokaryotic	Eukaryotic	Total
Glyceraldehyd-3-P DH	2	11	13
Ferredoxin	6	4	10
α-Amylase	3	4	7
Protease	5	0	5
Isocitrate DH	3	1	4
Aminopeptidase	0	3	3
Enolase	1	3	4
Aldolase	0	3	3
ATPase	1	1	2
Formyl FH_4 Synthetase	2	0	2
Glucose-6-P Isomerase	0	2	2
Methylene FH_4 DH	1	0	1

[a]*A detailed list of these proteins, and references citing their amino acid compositions, are available from the author.*

190

literature in terms of residue per mole and were then converted to residue per hundred residues to facilitate comparisons between proteins of differing molecular weights.
Because of the paucity of amide data for thermophilic amino
acid compositions, ASP/ASN and GLU/GLN were considered as
ASX and GLX. This was the data base used in subsequent
calculations.

An initial comparison was made between enzymes catalyz-
ing the same reaction. To do this, the mean amino acid composition of all the nonthermophilic proteins for a particular
enzymatic reaction was calculated, along with the standard
deviation for each amino acid mean. This composition was
then compared with that of the corresponding thermophilic
enzyme. Amino acids differing by more than three units of
standard deviation were considered to be significant. The
results of this rather crude test are summarized in Figure 1.
The first thing that is apparent is that differences do occur
in the amino acid compositions of thermophilic proteins; it
is also apparent, however, that there are no consistent patterns to these differences. It appears that ASX is always
decreased and that GLX is always increased in thermophiles.
There also appears to be a trend for hydrophobic amino acids
to be increased, although this point is debatable as seen
below. Also, it appears that if changes do occur in the
basic amino acids, they are increased. Finally, although it
is not shown in Figure 1, these differences are often quantitatively small, and therefore may result in only a few
residues per molecule being changed.

If one assumes that the changes noted in Figure 1 are
indeed related to the thermophilic nature of the proteins
involved, then two implications are apparent. First, the
hypothesis that proteins from thermophilic organisms do not
greatly differ from their nonthermophilic counterparts is
supported. Second, the variety of small and subtle changes
observed suggests that there may be a multiplicity of
mechanisms for stabilization of these proteins.

The thermophilic and nonthermophilic data pools were
then treated as two distinct classes and the mean amino acid
composition calculated for each. The results are shown in
Table 2 and are compared with the data of Reeck and Fisher
(1973) for 207 proteins from a variety of sources. The last
column gives the results of a standard Student's t-test performed on the thermophilic and nonthermophilic means. The
term, t_c, is the calculated value of t for the two means,

AMINO ACID DIFFERENCE IN THERMOPHILIC PROTEIN

ENZYME							
GPDH	GLY →	ALA ↑	LEU ↑	PHE →	ARG ↑		
Fdx	ASP →						
Amy	ASP →	SER →	GLU ↑	PRO ↑	GLY →	TRP →	
Amy	SER →	VAL →	LEU ↑				
Amy	ILE →	TRP ↑					
Prot	ILE ↑	TYR ↑	HIS ↑				
ICDH	ASP →	SER →	ILE ↑	HIS →			
API	GLY ↑	VAL ↑	ILE ↑	ARG ↑			
Enol	ASP →	SER ↑	GLU ↑	GLY ↑	ILE →	LYS →	ARG ↑
Enol	GLU ↑	LYS →	HIS →	ARG ↑			
Aldo	GLY ↑	TYR →					

Fig. 1. *The following enzyme abbreviations are used: GPDH: glyceraldehyde-3-phosphate dehydrogenase; Fdx: ferredoxin; Amy: α-amylase; Prot: protease; ICDH: isocitrate dehydrogenase; API: aminopeptidase; Enol: enolase; Aldo: aldolase.*

Table 2

Average Amino Acid Compositions of Thermophilic and Nonthermophilic Proteins

	Reeck and Fisher (207 proteins)		Non thermophilic (56 proteins)		Thermophilic (15 proteins)		t_c/t_0
	Mean	sd	Mean	sd	Mean	sd	
Asx	10.7	2.6	12.3	2.2	10.2	2.1	1.76
Thr	5.7	1.9	6.0	2.2	6.1	1.4	.14
Ser	6.3	2.5	6.3	1.6	4.4	1.4	1.62
Glx	10.6	3.3	9.0	2.9	10.2	2.6	.98
Pro	4.8	2.1	4.4	1.2	4.9	3.2	.42
Gly	8.1	3.1	8.7	1.8	9.3	1.4	.77
Ala	8.5	2.8	9.8	2.3	10.2	3.1	.33
Cys	2.3	2.7	2.9	4.3	1.6	4.0	.56
Val	6.8	2.0	7.6	2.1	8.0	2.0	.28
Met	1.9	1.1	1.6	.9	1.7	.8	.12
Ile	5.0	1.7	5.9	1.7	6.3	1.7	.25
Leu	8.1	2.5	6.4	2.6	7.1	2.6	.61
Tyr	3.3	1.6	3.3	1.6	3.5	2.4	.06
Phe	3.7	1.4	3.5	1.2	3.2	1.5	.54
Lys	6.5	2.7	5.9	2.6	5.9	1.5	.03
His	2.2	1.2	1.9	.9	2.3	.6	.12
Arg	4.4	2.0	3.1	1.4	3.8	1.4	1.07
Trp	1.3	1.0	1.2	1.4	1.2	1.5	.03

193

and is related to t_0, the theoretical value of t, at $p = 0.01$. Therefore, values greater than one indicate significant difference in the means. The data demonstrate significant differences occurring for ASX, SER, and ARG. At a confidence level of 95% ($p = 0.05$), GLX was also significantly different. However, these differences in the acidic amino acids (ASX and GLX) may perhaps be an artifact arising from the data pool. If one compares the thermophilic means with the data of Reeck and Fisher, no differences seem to occur.

The amino acid compositions were then used to calculate several structural parameters, which hopefully might give some insight into the capacity for structural interactions in these proteins. Krigbaum and Knutton (1973) have analyzed 18 proteins of known three-dimensional structure to determine the frequency of certain amino acids occurring in specific structural regions. From these observations, they developed equations to predict the helical, beta sheet, and turn content of an unknown protein from its amino acid composition. Estimates of hydrophobic capacity were made using the equation of Bull and Breese (1973), which is somewhat similar to the Bigelow (1967) average hydrophobicity.

Bull and Breese (1973) have related the hydrophobicity and the average residue volume to the melting point of a given protein. They experimentally measured the melting point of 14 proteins by following pH changes occurring upon heating unbuffered solutions of the protein. The observed melting point was then related by correlation analysis to the hydrophobicity (TM1), average residue volume (TM2), and both of these parameters (TM3). Bull and Breese noted an interesting relationship between the calculated melting points and the inactivation temperatures for enolases, as seen in Table 3. These data demonstrate that the predicted melting point for the thermophilic enolases is lower than that calculated for the nonthermophilic enzymes; there is an approximately inverse relationship between the temperature of one-half inactivation and the predicted melting temperature. We wished to see if this was a generalized observation for other thermophilic proteins.

The amino acid compositions for the proteins outlined in Table 1 were used to solve for the above parameters. Comparisons were then made between individual classes of enzymes. Statistical analyses were then performed, the results of which are summarized in Tables 4, 5, and 6. The ratio, t_c/t_0, is as already defined. The ratio, F_c/F_0, is the result of a

Table 3

Theoretical Melting Temperature
and Heat Inactivation for Enolase[a]

Source	$T_{1/2}$	T_{M1}	T_{M2}	T_{M3}
Muscle	40°	76	72	72
Yeast	45°	71	68	68
Thermus X-1	74°	67	69	70
T. aquaticus	90°	55	56	58

[a]*From Bull and Breese (1973). Arch. Biochem.
Biophys. 158, 681.*

variance ratio analysis, and provides a test for the validity
of using the Student's *t*-test. Values greater than 1.0 would
suggest that this test is not reliable.

Table 4 gives the results of solving the Krigbaum and
Knutton equations for structural parameters, and demonstrates
that no differences occur in the helical and turn content of
thermophilic relative to nonthermophilic proteins. The predicted beta content, however, is markedly lower for thermophilic proteins. This was found to be true for all of the
thermophilic proteins studied, except for the ATPase and the
formyltetrahydrofolate synthetase. Furthermore, these differences were often quite large, ranging as high as 25% difference between thermophilic and nonthermophilic proteins.

Table 5 gives the results of the hydrophobicity calculation, and suggests that no differences occur in this parameter for thermophilic proteins. A similar observation was
obtained with three other hydrophobicity parameters (Bigelow,
1967; Waugh, 1954; Fisher, 1964) for many of these proteins
(Singleton and Amelunxen, 1973).

Table 6 demonstrates that no significant difference was
observed for the various predicted melting temperatures for
thermophilic and nonthermophilic proteins.

Table 4

Structural Parameters for Thermophilic (T) and Nonthermophilic (NT) Proteins
Calculated by the Equations of Krigbaum and Knutton

	Per Cent Helix		Per Cent Turn		Per Cent Beta Sheet	
	T	NT	T	NT	T	NT
Mean	31.79	30.83	40.45	37.75	16.20	23.64
±1 SD	9.72	6.47	5.31	6.12	11.15	1.16
F_c/F_0	0.93		0.41		0.67	
t_c/t_0	0.34		0.71		1.63	

196

Table 5

Hydrophobicity Parameter of Bull and Breese
for Thermophilic (T) and Nonthermophilic (NT) Proteins

	HΦ	
	T	NT
Mean	968	936
±1 SD	85	71
F_c/F_0	0.37	
t_c/t_0	0.11	

Table 6

Predicted Melting Temperature
for Thermophilic (T) and Nonthermophilic (NT) Proteins

	TM1		TM2		TM3	
	T	NT	T	NT	T	NT
Mean	79.6	70.0	67.6	63.2	67.7	62.9
±1 SD	2.8	6.1	10.1	11.5	10.1	11.7
F_c/F_0	0.37		0.50		0.51	
t_c/t_0	0.10		0.31		0.31	

In summary, these studies demonstrate that differences
may occur in the amino acid composition of proteins from
thermophilic organisms, but these differences are quite
varied and very subtle. There does not appear to be any difference in the predicted values of helical content, turn
content, or hydrophobicity, but the beta sheet content of
thermophilic proteins may be decreased. It is not certain at

this time how this last observation may be related to the
thermostable nature of these proteins; however, Davidson and
Fasman (1967) suggested that perhaps beta structure might be
related to the thermal inactivation of proteins. The ratio-
nale behind this speculation was that beta structure might
lead to a rigid, relatively inactive protein, and that, at
least in the case of poly-lysine, beta structure is favored
by high temperature. From this, one might speculate that
perhaps thermophilic bacteria have adapted by modification of
protein sequences, such that beta structure formation is
avoided at high temperature. Instead of the increased hydrophobic interaction (at elevated temperatures) leading to beta
sheet formation, as was the case with poly-lysine, in thermophilic proteins such interactions may lead to softer, less
rigid structures, still capable of catalytic action.

Aside from these observations regarding beta structure,
there are other models of protein stabilization consistent
with the observations of minor and subtle differences reported
above. Thermophilic proteins might well be stabilized by
metal ions, such as has been demonstrated with thermolysin
(Matthews et al., 1972; Feder et al., 1971). Such a stabilization would require very minimal changes in the amino acid
composition of a protein. Furthermore, the binding of one or
more metal ions to a protein would cause relatively minor
changes in its molecular weight or other physical properties.
Few investigators have specifically examined their thermophilic systems for such an effect. Alternatively, a protein
might be stabilized via unique sequence features that rein-
force other structural aspects (e.g., helical content) of the
protein. Elsewhere in this volume, Dr. Reid demonstrated
that the stability of helical sequences of tRNA depend greatly
upon the bases at each end of the sequence. Extrapolating
from RNA to proteins, one can envisage helices and beta structures of differing stabilities, depending upon the specific
amino acids in the sequence. Again, this mechanism would
lead to very minor and subtle changes in amino acid composition for the protein.

It now seems safe to say that thermophilic bacteria survive by synthesizing proteins that are inherently capable of
withstanding the rigors of their environment. It is also
apparent that the primary distinguishing characteristic
between thermophilic proteins and nonthermophilic proteins
is this relative ability to withstand heat denaturation.
However, it is most likely that the mechanisms whereby these
proteins are stabilized, at a molecular level, are quite

subtle and may be manifold.

In retrospect, this conclusion should perhaps not be surprising. Thermal stability is possibly a more sensitive reflection of protein conformation than is enzymatic activity. Langridge (1968) and Ingraham (1973) have very elegantly demonstrated by genetic methods that, for β-galactosidase, changes in a single amino acid can cause marked loss of thermostability without concomitant losses in enzymatic activity. The literature of protein chemistry contains numerous similar examples in which the changing of a single amino acid in a protein sequence markedly changes the activity of the protein.

Thus, the thermophilic organism is most likely a very finely tuned result of many minor and subtle changes (from the structural, surely not the functional, point of view), which enable it to survive in its extreme environment.

ACKNOWLEDGMENT

The author wishes to acknowledge the support of a Research Associateship from the National Academy of Sciences, National Research Council. Present address: University of Delaware, Newark, DE 19711.

REFERENCES

BIGELOW, C. C. (1967). J. Theor. Biol. 16, 187.

BULL, H. B., and BREESE, K. (1973). Arch. Biochem. Biophys. 159, 681.

DAVIDSON, B., and FASMAN, G. D. (1967). Biochem. 6, 1616.

FEDER, J., GARRETT, L. R., and WILDI, B. S. (1971). Biochem. 10, 4552.

FISHER, H. F. (1964). Proc. Nat. Acad. Sci. U.S.A. 51, 1285.

INGRAHAM, J. L. (1973). In "Temperature and Life" (H. Precht, J. Christopherson, H. Hensel, and W. Larcher, eds.). Springer-Verlag, New York.

KRIGBAUM, W. R., and KNUTTON, S. P. (1973). Proc. Nat. Acad. Sci. U.S.A. 70, 2809.

LANGRIDGE, J. (1968). Mole. Gen. Genetics 103, 116.

MATTHEWS, B. W., COLMAN, P. M., JANSONIUS, J. N., TITANI, K., WALSH, K. A., and NEURATH, H. (1972). Nature (New Biol.) 238, 41.

REECK, G. R., and FISHER, L. (1973). Int. J. Peptide Protein Res. 5, 109.

SINGLETON, R., Jr., and AMELUNXEN, R. E. (1973). Bacteriol. Revs. 37, 320.

WAUGH, W. F. (1954). Advan. Prot. Chem. 9, 326.

KINETIC BEHAVIOR OF A THERMOPHILIC ENZYME IN RESPONSE TO TEMPERATURE PERTURBATIONS

C. R. Middaugh and R. D. MacElroy

INTRODUCTION

The observation that the proteins of thermophilic bacteria are generally more thermally stable than analogous proteins from mesophilic organisms (Singleton and Amelunxen, 1973) suggests that the two types of protein probably differ in one or more intrinsic properties. No doubt there are differences, but currently we are forced to consider any such variations as subtle, since most differences which have been found appear to bear no relationship to our tentative concepts of protein thermostability. We therefore must assume the position of amassing more information about the structure and behavior of various thermophilic and mesophilic proteins in the hope that differences which are, or will become, manifest will also become meaningful.

To NASA, the inhabitants of extreme environments are of interest because, from a geocentric viewpoint, conditions elsewhere in the solar system are "extreme." Furthermore, each discovery of an adaptation to an extreme environment on this planet improves our assessments of the ability of life to survive, and possibly to have evolved, in an extraterrestrial environment.

We would like to describe some work we have done with a thermophilic enzyme, ribose-5-phosphate isomerase (E.C. 5.3.1.6), from Bacillus caldolyticus and a mesophilic homo-logue, from the autotroph Thiobacillus thioparus. The mesophilic organism, T. thioparus, was grown as previously described by Santer et al. (1959). The growth of B. caldolyticus has been described previously (Weerkamp and MacElroy, 1972); it is an extreme thermophile and has a maximum growth temperature above 80°C. Cells were disrupted in a French press, and the enzymes were partially purified (Middaugh and MacElroy, 1976). The thermophilic isomerase was purified 275-fold, and the mesophilic enzyme about 90-fold.

The procedure of Axelrod and Jang (1954) was used to measure the formation of ribulose-5-phosphate. Incubations of enzyme and substrate at pH 7.5 in 0.1 M Tris with 1 mM each of Na_4EDTA and 2-mercaptoethanol were carried out at temperatures ranging from 5°C to 90°C. Substrate concentra-tions for kinetic investigations were varied between 0.1 and 12 mM. The response of the enzymes was linear when the data were plotted in the Lineweaver-Burk manner, and both enzymes appear to demonstrate simple Michaelis-Menten kinetics.

RESULTS

Both the mesophilic and the thermophilic enzymes have approximately the same molecular weight (40,000 daltons) and have essentially the same response to pH: Both are active between pH 6 and pH 10 and have a flat response between pH 7 and pH 9.

The temperature of maximum activity of the mesophilic enzyme was observed to be about 50°C while that of the ther-mophilic enzyme was about 65°C. When these data were plotted in the standard Arrhenius manner, discontinuities (T_D) were observed at 34°C with the mesophilic enzyme, and at 54°C with the thermophilic enzyme. The thermal stabilities of the enzymes were examined and it was found that after 1 hour at 57°C the mesophilic enzyme retained about 50% of its activity, while the thermophilic enzyme retained 50% of its activity after 1 hour at 91°C. The addition of a series of neutral salts or alcohols at 1.0 M concentration was found to alter the stability of both enzymes (Table 1). Ammonium sulfate, NaCl, KCl, and LiCl increased thermal stability, while n-propanol, $CaCl_2$, ethanol, methanol, and LiBr decreased it. Ethylene glycol increased the stability of the thermophilic enzyme, while it slightly decreased the stability of the

Table 1

The Effect of Additions on the Stability
of Ribose-5-Phosphate Isomerases

Additions (1.0 M)	At 57°C Mesophilic enzyme	At 91°C Thermophilic enzyme
None	50	50
$(NH_4)_2SO_4$	96	71
NaCl	79	59
KCl	75	57
LiCl	67	54
LiBr	33	31
$CaCl_2$	2	2
ethylene glycol	47	66
methanol	38	31
ethanol	26	5
n-propanol	0	0

mesophilic protein.

In view of the effect of these additives on stability, their effects on kinetic parameters were investigated. Using Lineweaver-Burk plots, the K_m and V_{max} for the enzymes were calculated at a number of temperatures in the presence and absence of neutral salts or alcohols. Plots for pK_m and pV_{max} versus $1/T$, both seemed best approximated by a pair of straight lines meeting at a rather well-defined transition temperature or point of discontinuity (T_D). The thermophilic enzyme exhibited a T_D at 53°C for K_m, and 35°C for V_{max}, while the mesophilic enzyme showed a T_D at 33°C for K_m, and at 11°C for V_{max}. The slopes of the data for the V_{max} plots are theoretically considered proportional to the energy of formation of the activated enzyme-substrate complex (Dixon and Webb, 1964). This value was calculated to be about 33,000 cal/mole at temperatures below the T_D (11°C) and about 17,000 cal/mole above the T_D for the mesophilic enzyme. The

corresponding values for the thermophilic enzyme were
36,000 cal/mole below 35°C and 17,000 cal/mole above.

The effect of neutral salts and alcohols at 1.0 M concentrations was very slight on the calculated activation
energies of either of the enzymes, although there appears to
be a tendency for these compounds to affect the thermophilic
enzymes more than the mesophilic. The effect of neutral
salts or alcohols on the rate of K_m change with temperature
was minimal at low temperatures in the case of both the mesophilic and the thermophilic enzyme. At high temperatures,
however, the effect of both salts and alcohols on the thermophilic enzyme was pronounced. Essentially, the rate of K_m
increase with temperature was the same as the control when
ethylene glycol, $CaCl_2$ or LiBr were present. The effect of
the other salts and alcohols was to decrease the rate of K_m
increase.

The most marked effect of the addition of neutral salts
and alcohols, however, was the displacement of the T_D's.
Table 2 summarizes these data. It can be seen that the temperature at the T_D observed in plots of V_{max} versus $1/T$ for
the thermophilic enzyme increased from 35°C up to 46°C in the
presence of $(NH_4)_2SO_4$. KCl, NaCl, ethylene glycol, and LiCl
were less effective in increasing the T_D; the other additives
decreased it. The mesophilic enzyme was similarly affected.
The T_D's indicated by the K_m data for both the mesophile and
the thermophile are similarly shifted by neutral salts and
alcohols.

DISCUSSION

Ribosephosphate isomerase, isolated from the extreme
thermophile Bacillus caldolyticus and the autotrophic mesophile Thiobacillus thioparus appears to have properties similar to the homologous enzyme from a variety of sources. Such
distinguishing characteristics as molecular weight, pH dependence, equilibrium ratio, activation energies (at low temperatures), and sensitivity to various inhibitors suggest few
distinguishing differences among the isomerases (Bruns et
al., 1958; Agosin and Arrevena, 1960; Matsushima and Simpson,
1965; David and Weismeyer, 1970; Noltman, 1972; Domagk et al.,
1973; Middaugh and MacElroy, 1976). The obvious exception
to this would seem to be the thermal properties of
the thermophilic enzyme. The enzyme from B. caldolyticus
displays both a higher temperature optimum and a greater
thermostability than the isomerase from mesophilic organisms.

Table 2

Temperature of Rate Change of K_m or V_{max} versus Temperature

Additions (1.0 M)	Thermophile		Mesophile	
	T_D V_{max}	T_D K_m	T_D V_{max}	T_D K_m
$(NH_4)_2SO_4$	46.2	60	17.7	39.5
KCl	39.9	57	15.2	37.4
NaCl	39.0	56.1	14.5	35.9
LiCl	36.8	54.1	12.7	35.7
LiBr	31.9	51.6	9.2	30.0
$CaCl_2$	28.2	49.6	6.8	27.7
None	35.2	53.4	11.1	32.6
ethylene glycol	38.23	56.59	13.28	34.88
methanol	31.69	51.46	8.92	32.43
ethanol	28.84	49.78	5.55	30.49
n-propanol	26.04	48.54	--	28.47

Furthermore, breaks (T_D) in the Arrhenius plots occur at
higher temperatures for the isomerase from the thermophile
than for the mesophile enzyme of this study.

Nonlinear relationships of the type observed in this
study, that seem best approximated by a pair of straight
lines meeting at some defined transition temperature, have
been rather commonly observed (Dixon and Webb, 1964; Massey
et al., 1966; Mazhul et al., 1970; Han, 1972; Drost-Hanson,
1972; Mavis and Bageles, 1972; Raison, 1973; Hachimuri and
Nosoh, 1973). These effects do not appear to be limited to
any particular type of molecular environment, having been
demonstrated for both monomeric and oligomeric proteins,
cytoplasmic and membrane-bound macromolecules, and from both
higher and lower plant and animal forms. A number of

explanations have been advanced to account for this observation (Dixon and Webb, 1964; Massey *et al.*, 1966; Mazhul *et al.*, 1970; Han, 1972; Drost-Hanson, 1972; Mavis and Bageles, 1972; Raison, 1973; Hachimuri and Nosoh, 1973). The hypothesis which has been most frequently postulated in the literature, and which at present seems most consistent with available experimental evidence, argues for the existence of temperature-dependent structural changes in either the subject macromolecule or its immediate environment. Studies of lipid-membrane associated enzymes have led some investigators to suggest that the liquid crystalline-to-crystalline phase change in the lipid component of the membranes was responsible for nonlinearity of Arrhenius plots (Epstein and Schechter, 1968; Raison, 1973). Drost-Hanson (1972) has proposed that changes in vicinal water structure could account for many reported thermal anomalies. Usually the conformational change is thought to occur directly in the structure of the macromolecule itself (Massey *et al.*, 1966; Mazhul *et al.*, 1970). In a number of studies, determinations of conformation-sensitive physical parameters as a function of temperature have strongly implicated "predenaturing" structural changes of this type (Massey *et al.*, 1966; Mazhul *et al.*, 1970; Hachimuri and Nosoh, 1973). This is consistent with recent suggestions for the existence of many macromolecules in multistable (metastable) conformational states (Nickerson, 1973) and with the occurrence of conformational isomers (conformers) (Epstein and Schechter, 1968) in a wide variety of biological situations. The preliminary evidence presented in this report suggests the most likely explanation for the temperature data for ribosephosphate isomerase from T. thioparus and B. caldolyticus is the presence of temperature-dependent conformational transitions of this general type. The differential effect of various inhibitory agents upon both enzymes above and below the transition temperatures can be explained most easily by assuming that different conformations of the enzyme exist in a temperature-dependent equilibrium. Perhaps most compelling, the shifts in critical transition temperatures (T_D) in the presence of alcohols and neutral salts are strikingly analogous to the effect of these solutes upon the well-characterized, conformational changes accompanying loss of enzymatic activity upon temperature denaturation for a variety of macromolecules (von Hippel and Schleich, 1969). Although an unambiguous demonstration of such a conformational change requires both purification of the isomerase to a homogeneous state and subsequent detailed physical studies, we shall neglect the less probable hypothesis that these temperature effects are due to temperature perturbations of ionization processes, or perhaps to the presence of nonconformational isozymes with differing energies of activation.

The question of the nature of the thermostability of macromolecules from thermophilic microorganisms may thus be approached by attempting to find conformational conversions between metastable states which occur at higher temperatures for the thermophilic protein than for its mesophilic counterpart. While we are aware of only one direct observation suggesting a conversion in a protein (Ljungdahl, this volume), the literature contains several examples of anomalous thermal behavior in homologous enzymes from mesophilic and thermophilic sources (Table 3). Intriguing evidence exists that such relationships may be detectable in temperature-dependent conformational transitions of transfer RNA molecules (Malcolm, 1969), as well as in ribosomes (Mednikov et al., 1972), and may perhaps be indirectly implicated in such complex forms of biological behavior as growth rates (Mednikov et al., 1972; Babel et al., 1972).

The hypothesis that the hydrophobic interactions may be a force in these transitions is attractive. A number of workers have discussed the possible existence of the phenomenon of "low temperature denaturation" (Brandts, 1969; Jarabak et al., 1966). Both theoretical and empirical considerations have implicated the unusual temperature dependence of the hydrophobic interaction in this process. If the predenaturing conformation transitions suggested in this paper are related to this low temperature structural lability, it may be hypothesized that the occurrence of these transitions at higher temperatures for thermophilic macromolecules is to some extent dependent on the hydrophobic interaction. Independent of the congruence of these concepts, which has yet to be established, the effect of a number of neutral salts and alcohols argues for the possibility of a functional and perhaps controlling role for apolar interactions in these effects. The effect of neutral salts upon macromolecular stability has not been clearly elucidated (von Hippel and Schleich, 1969), but most recent evidence suggests that the specificity of their action on proteins is due to the modulation of the binding of anions and cations to peptide amide dipoles by local nonpolar entities (Hamabata and von Hippel, 1973). The magnitude of these effects appears to be a function of such parameters as the relative position and hydrophobicity of the modulating apolar groups. The nonmediated binding of the salt itself appears to be relatively

Table 3

Thermal Anomalies in Enzymes from Thermophilic and Mesophilic Sources

Enzyme	Thermophilic source		Mesophilic source	
	T_D^a (°C)	Reference	T_D^a (°C)	Reference
Glucose-6-phosphate isomerase	55	(Muramatsu & Nosoh, 1971)	35	(Dyson & Noltman, 1968)
6-phosphogluconate dehydrogenase	30	(Miller & Shepherd, 1972)	19	(Miller & Shepherd 1972)
Formyltetrahydrofolate synthetase	40	(Brewer et al, 1970)	30	(Himes & Wilder, 1968)
ATPase	55	(Hachimuri & Nosoh, 1973)	18	(Charneck et al., 1971)
Pyrimidine ribonucleoside kinase	45	(Orengo & Saunders, 1972)	35	(Orengo & Saunders, 1972)
Glyceraldehyde-3-phosphate dehydrogenase	55	(Singleton & Amelunxen, 1973)	20	(Mazhul et al., 1970)
Fructose-1,6-diphosphate aldolase	52,60	(Sugimoto & Nosoh, 1971; Freeze & Brock, 1970)	30	(Sugimoto & Nosoh, 1971)
Ribosephosphate isomerase	54	This report	34	This report

a_{T_D}--approximate temperature of thermal anomaly.

nonspecific at high salt concentrations (Hamabata and von Hippel, 1973). The effect of various alcohols on both the transition temperature and thermostability for both iso-merases further hints at some ill-defined involvement of hydrophobic forces at higher temperatures for the thermo-philic enzyme (von Hippel and Schleich, 1969). It is inter-esting to note that studies of proteins from halophilic microorganisms have begun to implicate the hydrophobic inter-action in the salt-dependent stability of these macromole-cules by somewhat similar arguments (Lanyi and Stevenson, 1970).

It has been pointed out that there exists little corre-lation between the total hydrophobicity and thermostability of a protein, when comparisons are made between a thermo-philic protein and one or more of its mesophilic counter-parts (Singleton and Amelunxen, 1973). Thus, it may be hypo-thesized that the differences in transition temperature for the predenaturing conformational changes and perhaps for the mechanism of macromolecular thermal stability itself is due to the location of critically situated hydrophobic residues rather than to the total hydrophobicity of the protein (Singleton and Amelunxen, 1973).

The possible *in vivo* significance of these conforma-tional transitions is unclear. It is not difficult to ima-gine a role for this kind of conformational response in the adaptation of an organism to extremes in environmental tem-peratures (Hochachka and Somero, 1973), or in some type of integration into metabolic regulatory processes (Nickerson, 1973). The possibility that these effects are merely arti-factual manifestations of the inherent structural lability of macromolecules necessary for the normal *in vivo* function-ing of these molecules (Citri, 1973) cannot be excluded.

In summary, we would like to suggest the following: Macromolecules from a variety of sources appear to exist in a temperature-dependent equilibrium between two or more kine-tically distinguishable states within a physiological range of temperatures. Enzymes from thermophilic organisms seem to undergo these structural transitions at higher temperatures than their mesophilic counterparts. Speculative discussion has allowed for the possibility that the hydrophobic inter-action may be involved in these conformational changes, and may be further extended to suggest a possible involvment of this interaction in the marked temperature stability of pro-teins from thermophilic microorganisms. It must be

emphasized that the observations reported here are indirect
and require confirmation and interpretation beyond the scope
of this preliminary work.

REFERENCES

AGOSIN, M., and AREVENA, L. (1960). Enzymologia 22, 281.

AXELROD, B., and JANG, R. (1954). J. Biol. Chem. 209, 847.

BABEL, W., ROSENTHAL, H. A., and RAPOPORT, S. (1972). Acta.
        Biol. Med. Germ. 28, 565.

BRANDTS, J. F. (1969). In "Biological Macromolecules,"
        Vol. II (S. Timasheff and G. Fasman, eds.),
        pp. 213-290. Marcel Dekker, New York.

BREWER, J. M., LJUNGDAHL, L., SPENCER, T. E., and NEECE, S. H.
        (1970). J. Biol. Chem 245, 4798.

BRUNS, F. H., NOLTMANN, E., and VAHLHAUSE, E. (1958).
        Biochem. Z. 330, 483.

CHARNECK, J. S., COOK, D. A., and OPIT, L. J. (1971).
        Nature 233, 171.

CITRI, N. (1973). In "Advances in Enzymology," Vol. 37
        (A. Meister, ed.). Academic Press, New York.

DAVID, J., and WIESMEYER, H. (1970). Biochim. Biophys. Acta
        208, 56.

DIXON, M., and WEBB, E. C. (1964). "Enzymes," pp. 150-170.
        Academic Press, New York.

DOMAGK, G. F., DOERING, K. M., and CHILLA, R. (1973).
        Eur. J. Biochem. 38, 259.

DROST-HANSEN, W. (1972). In "Chemistry of the Cell Interface,"
        Part B (H. D. Brown, ed.), pp. 1-184. Academic
        Press, New York.

DYSON, J. E., and NOLTMANN, E. A. (1968). J. Biol. Chem.
        243, 1401.

EPSTEIN, C. J., and SCHECHTER, A. N. (1968). Ann. N. Y. Acad.
        Sci. 151, 85.

FREEZE, H., and BROCK, T. D. (1970). J. Bacteriol. 101, 541.

HACHIMURI, A., and NOSOH, Y. (1973). Biochim. Biophys. Acta 315, 481.

HAMABATA, A., and VON HIPPEL, P. H. (1973). Biochem. 12, 1264.

HAN, M. H. (1972). J. Theor. Biol. 35, 543.

HIMES, R. H., and WILDER, T. (1968). Arch. Biochem. Biophys. 124, 230.

HOCHACHKA, P. W., and SOMERO, G. N. (1973). "Strategies of Biochemical Adaptation," pp. 179-270. W. B. Saunders Co., Philadelphia.

JARABAK, J., SEEDS, Jr., A. E., and TALALAY, P. (1966). Biochem. 5, 1269.

LANYI, J. K., and STEVENSON, J. (1970). J. Biol. Chem. 245, 4074.

MALCOLM, N. L. (1969). Nature 221, 1031.

MASSEY, V., CURTI, B., and GANTHER, H. (1966). J. Biol. Chem. 241, 2347.

MATSUSHIMA, K., and SIMPSON, J. J. (1965). Canad. J. Microbiol. 11, 967.

MAVIS, R. D., and BAGELES, P. R. (1972). J. Biol. Chem. 247, 652.

MAZHUL, V. M., CHERNITSKII, Y. A., and KONEV, S. V. (1970). Biofizika 15, 5.

MEDNIKOV, B. M., SHUBINA, E. A., and TUROVA, T. P. (1972). Dokl. Akad. Nauk SSSR 205, 1240.

MIDDAUGH, C. R., and MacELROY, R. D. (1976). J. Biochem. (Japan), in press.

MILLER, H.M., and SHEPHERD, M.G. (1972). Canad. J. Microbiol. 18, 1289.

MURAMATSU, N., and NOSOH, Y. (1971). Arch. Biochem. Biophys. 144, 245.

NICKERSON, K.W. (1973). J. Theor. Biol. 40, 507.

211

NOLTMANN, E. A. (1972). In "The Enzymes," Vol. VI (P. D. Boyer, ed.), pp. 318-324. Academic Press, New York.

ORENGO, A., and SAUNDERS, G. F. (1972). Biochem. 11, 1761.

RAISON, J. K. (1973). Bioenergetics 4, 285.

SANTER, M., BOYER, J., and SANTER, U. (1959). J. Bacteriol. 78, 197.

SINGLETON, Jr., R., and AMELUNXEN, R. E. (1973). Bact. Rev. 37, 320.

SUGIMOTO, S., and NOSOH, Y. (1971). Biochim. Biophys. Acta 235, 210.

VON HIPPEL, P. H., and SCHLEICH, T. (1969). In "Biological Macromolecules," Vol. II (S. Timasheff and G. Fasman, eds.), pp. 417-574. Marcel Dekker, New York.

WEERKAMP, A., and MacELROY, R. D. (1972). Arch. Mikrobiol. 85, 113.

PROPERTIES OF A PARTIALLY PURIFIED NADH DEHYDROGENASE FROM AN EXTREMELY HALOPHILIC BACTERIUM, STRAIN AR-1

Lawrence I. Hochstein

The extremely halophilic bacteria appear to have acquired several properties not usually associated with the more mundane microbes that are a microbiologist's usual experience. Among these properties is the presence of a high intracellular salt concentration. For example, one such extremely halophilic bacterium, Halobacterium salinarium, has been reported to have an internal sodium concentration of 1.37 molal, whereas the potassium concentration was reported to be 4.5 molal (Christian and Waltho, 1962).

The presence of this high internal salt concentration would be expected to affect the enzymatic machinery of the cell, and early studies on the enzymes from the extremely halophilic bacteria confirmed this assumption (Baxter and Gibbons, 1956; Larsen, 1967). An example of the peculiar behavior of enzymes from extremely halophilic bacteria is shown in Figure 1, which describes the effect of NaCl on the activity of malic dehydrogenase activity in crude extracts from H. salinarium, strain 1 (Hochstein, unpublished data). The enzyme was activated by NaCl, with maximum activity occurring at about 1 M NaCl. Higher concentrations of NaCl, however, inhibited the enzyme so that at a concentration of 4 M NaCl the activity was about 55% of maximum. Considerable

THE EFFECT OF NaCl ON MDH ACTIVITY

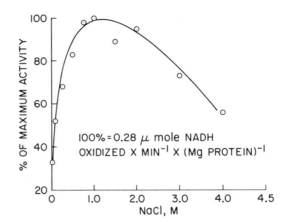

Fig. 1. *The effect of NaCl on the activity of malic dehydro-*
genase from Halobacterium salinarium, *strain 1.*
Crude extracts were assayed in reaction mixtures
containing the following additions in a total volume
of 1 ml: Tris-HCl, pH 7.8 (50 µmoles); oxaloacetic
acid (0.3 µmoles); NADH (0.1 µmole); NaCl, as indi-
cated. The reaction was carried out at 22°C and
initiated by the addition of enzyme.

activity was detected in the absence of added NaCl (NaCl $\cong$
23 mM). Extrapolation of the data suggested that the enzyme
exhibited about 20% of the maximum activity in the absence
of NaCl. Substituting KCl for NaCl did not materially alter
the results.

An additional property characteristic of halophilic
enzymes is the marked dependency of enzyme stability on ionic
strength (Larsen, 1967). This effect is shown in Table 1,
which describes the effect of various concentrations of NaCl
on the stability of malic dehydrogenase activity in crude
extracts from H. salinarium, strain 1 (Hochstein, unpublished
data). In the presence of 23 mM NaCl (a concentration where
enzyme activity was 35% of maximum), approximately 3.8% of
the enzyme was inactivated per second. Even in the presence
of 1 M NaCl, the enzyme was unstable, losing about 0.02% of

Table 1

The Stability of Malic Dehydrogenase
from Halobacterium salinarium, Strain 1,
as a Function of Sodium Chloride Concentration[a]

NaCl M	Inactivation constants $K(sec^{-1})$
0.023	0.038
0.125	0.012
0.250	0.0046
0.5	0.0008
1.0	0.00019

[a]*Crude extracts prepared from Halobacterium salinarium were diluted in 50 mM Tris-HCl buffer, pH 7.8, to the desired final concentration of sodium chloride. In all cases dilutions were made such that the final concentration of crude extract was the same. Aliquots were removed at different times and assayed for residual malic dehydrogenase activity in reaction mixtures containing the following additions in a total volume of 1 ml: 50 μmoles Tris-HCl, pH 7.8; 0.3 μmoles oxaloacetate; and 0.1 μmole NADH. The initial rate of NADH oxidation was taken as a measure of the residual activity, and semilog plots of enzyme activity versus time were constructed. The first order inactivation constants were calculated from these plots.*

its activity per second. Replacing NaCl with KCl had no appreciable effect on the stability of the enzyme.

Our investigations into the effect of salts on the activity of halophilic enzymes were initiated while studying a membrane-bound NADH oxidase from an extremely halophilic bacterium isolated from a local saltern (Hochstein and Dalton, 1968). The oxidation of NADH was enhanced by various monovalent cations. However, the efficacy of a cation was related to the accompanying anion. Thus while LiCl failed to stimulate NADH oxidase activity, Li_2SO_4 proved extremely

effective. A similar cation-anion relationship was observed
with magnesium ions. $MgCl_2$ stimulated NADH oxidase activity
at low concentrations; concentrations greater than 100 mM
were inhibitory. On the other hand, $MgSO_4$ enhanced NADH oxidation up to concentrations as high as 2 M. The importance
of anions in the cation activation of NADH oxidase activity
was further confirmed by observing that the oxidation of
NADH, in the presence of various sodium salts, attained a
maximum value at lower salt concentrations in the presence of
citrate, phosphate, and sulfate, as compared to chloride.
Enzyme activity was either inhibited or not stimulated by
chaotropic ions such as nitrate, bromide, or iodide.

These results were not readily reconciled with the hypothesis that monovalent cations acted by charge neutralization
(Baxter, 1959), but rather suggested that the stabilization
of hydrophobic interactions was of some importance (Lanyi and
Stevenson, 1970). In order to better study these ionic effects in a system uncomplicated by the presence of a multi-
enzyme system, particularly one enriched with respect to the
hydrophobic components associated with the bacterial membrane, the NADH dehydrogenase activity associated with the
membrane fraction was chosen for further study. This enzyme
was selected since previous experiments suggested that many
of the ionic effects observed with the NADH oxidase system
were associated with the NADH dehydrogenase.

Several properties facilitated the purification of the
NADH dehydrogenase (Hochstein and Dalton, 1973). The enzyme
was readily solubilized from the membrane in the presence of
2 M NaCl, and it was stable at this salt concentration.
Furthermore, although the enzyme was unstable in solutions
of low ionic strength, NADH markedly retarded the inactivation process. Finally, after inactivation in solutions of
low ionic strength, the enzyme could be reactivated following
the direct addition of NaCl. The purification of the enzyme
was initiated by passing crude extracts through an Agarose
0.5m column equilibrated in 2 M NaCl. This resulted in the
solubilization and separation of NADH dehydrogenase activity
from the membrane fraction. Such preparations were free of
cytochrome components as evidenced by the failure to detect
Soret bands following reduction with dithionite. Furthermore, the process appeared to be fairly selective since the
membrane fraction, following gel filtration in 2 M NaCl, was
still capable of oxidizing α-glycerophosphate, as well as
ascorbate and TPMD-ascorbate (Hochstein, unpublished data),
both of which have been shown to introduce electrons at the

cytochome level in membrane preparations obtained from extremely halophilic bacteria (Cheah, 1970).

The NADH dehydrogenase was further purified (Hochstein and Dalton, 1973) by hydroxylapatite chromatography followed by chromatography using QAE-Sephadex. Following reactivation of the enzyme, about 30% of the initial activity was reco-vered. The final product was purified some 438-fold over the starting material and, when assayed at fixed nonsaturating concentrations of DCIP and NADH, oxidized 241 μmoles of NADH/min/mg of protein. The results of a typical purifica-tion are shown in Table 2.

Table 2

The Purifcation of NADH Dehydrogenase
from the Extremely Halophilic Bacterium, Strain AR-1[a]

Fraction	Total units	Total protein (mg)	Specific activity	Purif.	Recovery (%)
Crude	880	1600	0.55	1.0	100
Agarose	675	210	3.2	5.8	77
Hydroxylapatite	400	23	17.4	32	45
QAE-Sephadex	265	1.1	241	438	30

[a]A unit of NADH dehydrogenase activity is defined as the amount of enzyme causing the reduction of 1 μmole of DCIP per minute. Specific Activity is units per mg protein.

As shown in Table 3, the enzyme oxidized a limited num-ber of NADH analogs. NADHP, deamino NADH, and 3-pyridine aldehyde-NADH were not oxidized. Substitution in the three position of the pyridine ring to give either the thioamide or the acetyl derivatives of NADH resulted in analogs which served as substrates, although they were not as effective as

217

Table 3

The Substrate Specificity of the NADH Dehydrogenase

Substrate	Activity
NADH	+
3-Acetylpyridine deamino NADH	+
Thionicotinamide-NADH	+
3-Acetylpyridine-NADH	+
Deamino-NADH	−
3-Pyridine aldehyde NADH	−
NADPH	−

NADH when judged by the maximum velocity at saturating concentration of substrate (Hochstein and Dalton, 1973).

In order to ascertain whether the inactive analogs failed to bind to the enzyme, NADH oxidation was carried out in the presence of the inactive substrates. Of those, only the 3-pyridine aldehyde derivative affected NADH oxidation (Table 4). The oxidation of NADH was competitively inhibited by the analog, suggesting that it was competing with NADH for the same binding site. These analog experiments suggested that NADH was bound to the enzyme via the amino-nitrogen of the purine ring since only those compounds having a free 6-amino nitrogen in their adenosine moieties were either active as substrates or acted as competitive inhibitors.

In order to confirm this hypothesis, an experiment was carried out in which the oxidation of NADH was determined in the presence of NAD, AMP, or nicotinamide mononucleotide. As indicated in Table 4, only NAD and AMP, both possessing a purine amino-nitrogen, inhibited the enzyme. The inhibitions were competitive, with identical inhibition constants. Nicotinamide mononucleotide had no effect on NADH dehydrogenase activity.

Kinetic and product inhibition studies indicated that an ordered bi-bi mechanism (Plowman, 1972) most adequately

218

Table 4

Inhibition of NADH Dehydrogenase Activity[a]

Addition	Type of inhibition	K_i (mM)
3-Pyridine aldehyde NADH	C	0.4
Deamino-NADH	N	--
NAD	C	1.5
AMP	C	1.5
Nicotinamide mononucleotide	N	--

[a]*The NADH dehydrogenase was assayed in standard reaction mixtures in which the DCIP concentration was held constant, whereas the concentrations of NADH and NADH analog were varied. C represents competitive inhibition, N represents no inhibition, and K_i represents the inhibition constant for the particular inhibitor.*

described the kinetic properties of the NADH dehydrogenase. From these studies it was possible to identify NADH as the first substrate to add, and NAD as the last product to leave the enzyme.

As shown in Table 5, NADH dehydrogenase was activated by NaCl. Considerable activity was observed in the absence of added NaCl, amounting to some 59% of the maximum activity. Since the enzyme was assayed at dye and substrate concentrations not too far removed from their apparent K_m values, it was possible that the NaCl stimulation of NADH dehydrogenase activity was related to the limiting concentration of substrate rather than enzyme activation. This possibility was further investigated by carrying out a series of experiments in which enzyme activity was determined at various concentrations of both substrates. The results of these studies are shown in Figure 2. When NADH dehydrogenase activity was determined as the maximum velocity, the dependency of enzyme activity on salt concentration disappeared. Of the kinetic parameters measured, only K_{DCIP} (the apparent Michaelis constant for DCIP) changed, so that as the salt concentration increased, the affinity of the enzyme for the dye decreased.

Table 5

The Effect of Sodium Chloride on NADH Dehydrogenase Activity[a]

NaCl (M)	$-\Delta A_{600}$/min.
0	0.428
0.25	0.473
0.50	0.480
1.0	0.593
2.0	0.720
2.5	0.720

[a]*The enzyme was assayed at the indicated concentration of NaCl in reaction mixtures containing the following additions in a total volume of 1 ml: 50 μM Imidazole, pH 7.0; 0.1 μmole NADH; 70 μmoles DCIP; 1 μmole KCN. The reaction was initiated by the addition of enzyme.*

THE EFFECT OF NaCl ON THE KINETIC PARAMETERS
OF THE NADH-DEHYDROGENASE

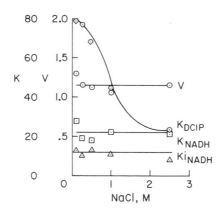

Fig. 2. *The effect of NaCl on the kinetic properties of the NADH Dehydrogenase. The data were obtained from double reciprocal plots in which the observed velocities were determined at various changing concentrations of one substrate in the presence of fixed concentra-tions of the other. The symbols represent (o), V max; (◇), K_{DCIP}; (□) K_{NADH}; (△), K_{iNADH}.*

If it is assumed that K_{DCIP} is the dissociation constant for this dye, then these results suggest that the apparent activation of the NADH dehydrogenase by NaCl is due to assaying the enzyme at the limiting concentration of DCIP.

Since high salt concentrations would be expected to minimize ionic interactions involved in the binding of substrates to the enzyme, these results are more easily explainable in terms of the ability of NaCl to neutralize electrostatic interactions so as to allow the dye, in an anionic configuration, to approach its binding site. If so, then the cation specificity should not be pronounced. This indeed turned out to be the case, as shown in Table 6. In addition, the values for V_{max} and K_{NADH} observed at low cation concentrations (0.1 M) were similar to those at high salt concentration. The only parameter affected was K_{DCIP}, the values of which while higher, were virtually identical for these

Table 6

The Cation Specificity of NADH Dehydrogenase:
The Effect on the Kinetic Parameters[a]

Cations	Conc. (M)	V	K_{NADH}	K_{DCIP}
LiCl	2.5	1.01	19.6	27.2
NaCl	2.5	1.14	25.4	28.0
KCl	2.5	1.09	17.9	16.6
CsCl	2.5	1.18	29.7	27.7
LiCl	0.1	1.32	35.2	77.9
NaCl	0.1	1.30	27.1	79.5
KCl	0.1	1.11	23.5	75.5

[a]*NADH dehydrogenase activity was determined in the presence of the indicated salts by measuring the observed velocities at various changing concentrations of one substrate in the presence of fixed concentrations of the other. The constants were obtained from double reciprocal plots and the secondary plots derived from them.*

different cations.

   Since earlier studies (Hochstein and Dalton, 1968) indicated that cation activation of NADH oxidase activity was
dependent on the nature of the accompanying anion, experiments were carried out to determine the effect of anions on
NADH dehydrogenase activity. As shown in Tables 6 and 7, the
activation of NADH dehydrogenase by sodium ion was dependent
on the nature of the anion. The activity of the enzyme followed the series $SO_4 > Cl > NO_3$, with thiocyanate inhibitory
at all the concentrations tested. The effect of the anions
on the kinetic parameters are presented in Table 8. The
dependency of enzyme activity on the nature of the anion disappeared when activity was determined at saturating concentrations of both substrates. Of the kinetic parameters measured, only K_{NADH} was affected, the magnitude of the constant
following the same order as observed for the observed velocity changes. Since the substrate specificity studies suggested that NADH was bound to the enzyme through its purine
amino nitrogen, it seemed unlikely that anions functioned by
neutralizing positive charges. The relationship between

Table 7

The Effect of Anions on the Activity of NADH Dehydrogenase[a]

Na$^+$ (M)	Observed velocity		
	SO_4^{2-}	NO_3^-	SCN^-
0	0.414	0.375	0.382
0.25	0.540	0.473	0.323
0.5	0.615	0.525	0.240
1.0	0.795	0.608	0.158
2.0	0.875	0.615	0.015
2.5	0.935	0.630	0

[a]*NADH dehydrogenase activity was determined in the standard reaction mixtures in which NaCl was replaced by the
indicated salts.*

Table 8

The Effect of Anions on the Kinetic Parameters
of NADH Dehydrogenase[a]

Anion	V	K_{NADH} (µM)	K_{DCIP} (µM)	K_{iNADH} (µM)
SO_4^{2-}	1.26	10.3	22.8	3.5
NO_3^-	1.12	50.8	28.6	14.2

[a]*NADH dehydrogenase activity was determined in the presence of various anions at a sodium ion concentration of 2.5 M. The observed velocities were obtained at various changing concentrations of one substrate in the presence of fixed concentrations of the other. The various kinetic parameters were obtained from double reciprocal plots and the secondary plots derived from them.*

observed velocity, K_{NADH}, and the lyotropic order of the anions suggested that the anion effects were associated with changes in the conformation of the enzyme in a way that affected the accessibility of NADH to its binding site.

This interpretation would be consistent with other observations that the conformation of several enzymes from extremely halophilic bacteria was affected by the nature of the anion. In particular, it was observed that salting-in type anions (e.g., sulfate) maximized hydrophobic interactions, leading to relatively compact forms of the enzyme, whereas salting out type anions (e.g., nitrate) tended to minimize these interactions and lead to less compact enzyme forms (Lanyi, 1974).

Thus, the observations on the effect of salts on NADH dehydrogenase activity can be interpreted as the result of two effects. Cations act by charge neutralization of negative charges about the dye binding site so as to minimize charges that prevent the approach of the dye to its binding site. Anions, on the other hand, affect the conformation of the enzyme so as to make the NADH binding site more

223

accessible to the bulk solvent and, hence, NADH. Ions such
as sulfate, which result in a compact enzyme form, presumably
lead to the maximum exposure of the NADH binding site, whereas ions such as nitrate result in a less compact form of the
enzyme in which the substrate binding site is relatively far
removed from the bulk solvent. These effects are reflected
in the increased apparent Michaelis constant for NADH. In
addition, these results indicate that if NADH dehydrogenase
activity is measured under conditions ensuring saturation by
both substrates, the dependency of enzyme activity on salt
concentration would disappear. Thus, the uniqueness of at
least this halophilic enzyme is not that it requires salt for
activity, but that salts interact with the enzyme (either by
charge neutralization or by maintenance of optimal conformational states) so as to enhance substrate binding.

The generality of these observations is uncertain at the
present time since only a few halophilic enzymes have been
studied in sufficient detail to allow for an adequate analysis. It may be pointed out that one such enzyme, a highly
purified alkaline phosphatase from H. salinarium (McParland,
1969) was shown to respond to salts in a manner similar to
that described here for the NADH dehydrogenase. Preliminary
studies with the NADH dehydrogenases from several other
extremely halophilic bacteria suggest that a similar picture
may emerge. Interestingly, the NADH dehydrogenase from
H. salinarium does not behave this way (Hochstein, unpublished
data). It is interesting that in every case so far studied,
those extremely halophilic bacteria that were activated by
salts as described in this paper were those that:
(1) utilize carbohydrates; (2) possess NADH dehydrogenases
that are readily solubilized from their membranes by 2 *M* NaCl,
and (3) are stable at that salt concentration.

REFERENCES

BAXTER, R. M., and GIBBONS, N. E. (1956). Canad. J. Microbiol.
2, 599.

BAXTER, R. M. (1959). Canad. J. Microbiol. 5, 47.

CHEAH, K. S. (1970). Biochim. Biophys. Acta 216, 43.

CHRISTIAN, J. H. B., and WALTHO, J. A. (1962). Biochim.
Biophys. Acta 65, 506.

HOCHSTEIN, L. I., and DALTON, B. P. (1968). J. Bacteriol. 95,
37.

HOCHSTEIN, L. I., and DALTON, B.P. (1973). *Biochim. Biophys. Acta* 302, 216.

LARSEN, H. (1967). *In* "The Bacteria: A Treatise On Structure And Function" Vol. IV, (I. C. Gunsalus, ed.), pp. 297-342. Academic Press, New York.

LANYI, J. K., and STEVENSON, M. J. (1970). *J. Biol. Chem.* 245, 4074.

LANYI, J. K. (1974). *Bacteriol. Rev.* 38, 272.

McPARLAND, R. H. (1969). Dissertation, University Microfilms, Ann Arbor, Michigan.

PLOWMAN, D. M. (1972). "Enzyme Kinetics." McGraw-Hill Book Co., New York.

MEMBRANES

EFFECT OF TEMPERATURE ON MEMBRANE PROTEINS

Neil E. Welker

The growth of microorganisms at relatively high temperatures has posed the question: How do these organisms manage to live, and in many instances, require temperatures where ordinary proteins and other cellular constituents are inactivated or denatured? A variety of theories have been proposed, but of these, only two have received experimental support. The first and most obvious is that the essential cell components of thermophilic organisms are relatively more heat stable than those of their mesophilic counterparts. The second theory considers thermophily to be due to an enhanced integrity of the membrane, and attempts to correlate the heat stability of the organism with the melting point of the membrane lipid. There is considerable evidence that pure enzymes and other proteins of thermophilic organisms are relatively more resistant to thermal denaturation than are their counterparts isolated from mesophilic organisms.

A majority of the studies on the effects of temperature on microbial cell constituents have been done with cell extracts, or, when possible, with purified preparations of the constituent. It is likely, however, that temperature has quite different effects on cell constituents *in vitro* as compared with *in vivo*. So far, very few studies have been reported which might throw light on possible differences in the temperature sensitivity of biological molecules *in vivo* and *in vitro*. It has been reported that substrates and coenzymes (Burton, 1951) can protect certain enzymes against thermal denaturation. Boyer *et al.* (1946, 1947) reported that short- and medium-chain fatty acids can protect some proteins against thermal denaturation.

Any consideration of the mechanisms of microbial adaptation to extreme environments must include detailed studies of isolated cellular constituents as well as the organization of these constituents in the cytoplasmic membrane.

Singer (1971) and Singer and Nicolson (1972) have placed membrane proteins into two general classes, which have been designated as peripheral and integral. Peripheral proteins are probably loosely bound to the membrane by electrostatic or hydrophobic interactions, whereas integral proteins are partially embedded in and partially protruding from the membrane. A majority of the protein in most membranes can be designated as integral. A schematic representation of a cross section of the lipid-globular protein mosaic model of membrane structure (Singer and Nicolson, 1972) is shown in Figure 1. The extent to which integral proteins are embedded in the membrane depends on their size and structure.

In this contribution, I will discuss only those reports that deal with the effect of temperature on enzymes

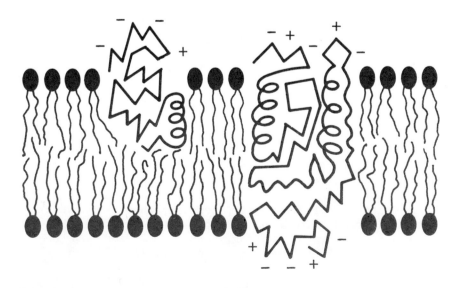

associated with the membrane (integral proteins). These studies preclude the use of purified proteins, but require intact cells or cell extracts in which the enzymes under investigation are attached to either the membrane or a membrane fragment or membrane lipid.

Indirect evidence that the temperature sensitivity of some enzymes may be different *in vivo* as compared with *in vitro* has come from a number of different sources. Eidlic and Neidhardt (1965) isolated a temperature-sensitive mutant of <u>Escherichia</u> <u>coli</u> which grew normally at 30°C, but failed to grow at 37°C; the parental strain grows optimally at 37°C. The temperature-sensitive mutant possessed an altered Valyl-tRNA synthetase which was active *in vivo* at 30°C but not at 37°C. In cell-free extracts, however, the temperature-sensitive enzyme was not active at 30°C.

Bubela and Holdsworth (1966) compared the thermostability of the amino acid activating system of membranes of <u>Bacillus</u> <u>stearothermophilus</u> with a soluble activating system prepared from the membranes. Membranes lost only 20% of their activity for the incorporation of amino acids into proteins after exposure to a temperature of 63°C for 10 min, whereas the detached system lost 50% of its activity after 10 min at 63°C (Figure 2).

Ameluxen and Lins (1968) compared the thermostability of 11 different enzymes in crude extracts of a thermophile and a mesophile. With the exception of pyruvate kinase and glutamic-oxaloacetic transaminase, the thermophile enzymes showed a much greater thermostability than those of the mesophile (Table 1). Above 60°C, pyruvate kinase and the glutamic-oxaloacetic transaminase from the thermophile are inactivated to the same extent as the corresponding enzymes from the mesophile. Although the two thermophile enzymes are inactivated *in vitro* at 70°C, the thermophile is capable of growth and division at these temperatures.

These combined results, although of a preliminary nature, indicate that the nature of the association of the

---

Fig. 1. The lipid-globular protein mosaic model of membrane structure. From Singer (1971).

231

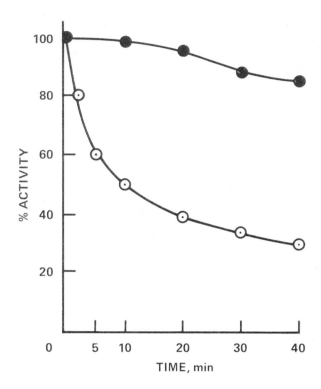

Fig. 2.   Heat stability of amino-activating systems of
         B. stearothermophilus held at 63°C. Protoplast
         membranes, ○ ; Sonicated membranes, ● . From
         Bubela and Holdsworth (1966).

enzyme with the membrane may contribute to the thermostability
of the enzyme. There are, of course, other interpretations,
such as the presence of specific or nonspecific protective
components in the crude extracts used in most of these experiments. It is also possible that there is a rapid turnover of
enzymes during growth at elevated temperatures. Unfortunately, no detailed studies were made on the membrane-bound and
detached enzymes, so these alternative interpretations cannot
be eliminated.

As suggested by Singleton and Amelunxen (1973), the

Table 1

Heat Inactivation of Pyruvate Kinase
and Glutamic-oxaloacetic Transaminase from
Bacillus stearothermophilus and Bacillus cereus[a]

Enzyme	Percentage of Inactivation After 10 Min					
	60°C		70°C		80°C	
	T^b	M^c	T^b	M^c	T^b	M^c
Pyruvate Kinase	4	34	99.9	99.9	100	100
Glutamic-Oxalo-acetic Transaminase	0	57	99.4	100	100	100

[a]From Amelunxen and Lins (1968).

[b]T = thermophile.

[c]M = mesophile.

scarcity of reports of this nature may reflect the hesitancy
of investigators to publish data on the thermolabile enzymes
that have been isolated from thermophiles.

The protective association of an enzyme with the membrane is a fundamental property of protein-lipid, protein-
protein, and protein-substrate interactions, and is probably
not unique to thermophilic organisms.

A general approach that has proven to be extremely useful in studies involving the effects of temperature on biological systems was first used by Crozier and collaborators in
the early 1920s (Ingraham, 1962). They studied the effect of
temperature on a large number of varied biological processes.
They clearly established that for most biological processes
there is a considerable range of temperature over which the
temperature-activity relation closely follows the Arrhenius
equation. This general approach has been used successfully
to examine the temperature-dependent activity of a function
of the membrane (membrane-bound enzymes) in response to
changes in the lipid composition of the membrane phospholipid

The utility of this approach depends upon the controlled manipulation of the composition of the membrane components.

Acholeplasma laidlawii is particularly suited for these studies because the fatty acid composition of the membrane lipids can be manipulated by varying the fatty acid supplement which is added to the growth medium (McElhaney and Tourtellotte, 1970). The absence of a cell wall also makes this organism suitable for studies on lipid phase transition using differential scanning calorimetry (Steim, 1972; Steim et al., 1969; DeKruyff et al., 1972; Melchior et al., 1970) and x-ray diffraction (Engelman, 1970). From these studies it was established that the fatty acids of membrane phospholipids undergo a transition from an ordered crystalline or gel state to a more disordered liquid-crystalline state between 20°C and 30°C.

DeKruyff and collaborators (1973) investigated the membrane-bound NADH oxidase system, p-nitrophenylphosphatase, and Mg^{2+}-dependent ATPase activity in A. laidlawii membranes as a function of temperature and fatty acid composition of the membrane phospholipids. The fatty acid composition of A. laidlawii membrane lipids of cells grown in media containing various fatty acids is shown in Table 2. The fatty acid supplement added to the growth medium dramatically increases its incorporation into the membrane, as well as causing minor fluctuations in the concentration of the other membrane lipid fatty acids. Arrhenius plots of the NADH oxidase system and p-nitrophenylphosphatase activities showed no discontinuities, although the membranes underwent a phase transition within the temperature range of 5°-35°C (Table 3). Membranes enriched in oleic (18:1 C) or linoleic (18:2 C) acid show no phase transition within this temperature range. In contrast, Arrhenius plots of the ATPase activity in membrane enriched with elaidic, stearic, or palmitic acids show breaks which roughly coincide with the temperature at which the membrane lipid phase transition begins (Table 3). DeKruyff et al. (1973) concluded that a membrane lipid phase transition induces a change in the activation energy of membrane ATPase. NADH oxidase and p-nitrophenylphosphatase, although membrane-bound, are not as sensitive to changes in the fatty acid composition of the membrane lipid.

A number of other studies using A. laidlawii or fatty acid auxotrophic mutants of E. coli support the theory that the liquid-crystalline state of the membrane lipids is necessary to support growth (Overath et al., 1970; McElhaney,

Table 2

Fatty Acid Composition of A. laidlawii Membrane Lipids of Cells Grown in Media Supplemented with Various Fatty Acids[a]

Fatty acid supplemented in the growth medium	Fatty acid composition of the membrane lipids (mole%)										
	12:0	13:0	14:0	15:0	16:0	17:0	18:0	18:1c	18:1t	18:2	Unknown
18:1t	4.9	0.2	3.0	0.9	14.0	--	1.8	--	69.8	1.1	4.3
18:0	20.1	2.2	21.2	1.4	16.4	0.4	30.9	1.0	--	1.1	5.3
16:0	7.6	1.2	20.2	0.9	62.7	--	1.7	2.8	--	1.5	1.5
18:1c	5.8	1.1	6.7	1.5	26.4	0.3	6.5	45.1	--	1.4	5.2
18:2	3.5	Trace	7.9	Trace	45.7	Trace	1.9	3.6	--	33.2	4.2

[a]From DeKruyff et al. (1973).

235

Table 3

Correlation Between the Endothermic Lipid Phase Transition
of A. laidlawii Membrane and the Temperature of the Break
in the Arrhenius Plot of ATPase Activity[a]

Fatty acid supplement in growth medium	Mole per cent in membrane lipid	Temperature at which phase transition begins (°C)	Break in the Arrhenius plot of ATPase activity (°C)
Elaidic (18:1t)	69.8	13	15
Stearic (18:0)	30.9	20	18.5
Palmitic (16:0)	62.7	25	18

[a]From DeKruyff et al. (1973).

1974), membrane transport (Overath *et al.*, 1970; Machtiger
and Fox, 1973; Wilson and Fox, 1971), respiration (Overath
et al., 1970), and the activity of membrane-associated enzymes (Kimelberg and Papahadjopoulos, 1972). Arrhenius plots
for most of these processes were biphasic, the slopes extrapolating to intersections at unique transition temperatures.
The transition temperatures for most of these biological processes varied with the degree of unsaturation of the predominant fatty acid in the membrane phospholipid.

In most studies of membrane-bound enzymes the temperature range was not extended beyond 40°C or 42°C. It would
have been interesting if data were available to indicate
what effect the fatty acid composition of the membrane lipids
had on the high temperature inactivation (thermal denaturation) of the various enzymes. No data, however, were presented as to the membrane content of the cell or the protein
content of the various membrane preparations.

Studies such as those done with the fatty acid auxotrophic mutants of E. coli or A. laidlawii have not been done
with thermophilic organisms. This is not surprising since
fatty acids of the membrane lipids of thermophiles cannot
easily be manipulated in the same manner. Changes in the
membrane content of the cell or quantitative and qualitative

changes in membrane lipid and proteins of thermophiles can be accomplished, however, by raising the temperature of growth.

Studies on membranes of thermophilic organisms have concentrated mainly on the fatty acids of the membrane lipid. The fatty acid distribution in mesophilic and thermophilic strains of the genus Bacillus (Cho and Salton, 1964; Shen *et al.*, 1970) and in psychrophilic, mesophilic, and thermophilic strains of the genus Clostridium (Chan *et al.*, 1971) have been determined. The combined data indicate that membranes of thermophiles generally contain a higher content of saturated and branched-chain fatty acids. In obligate (Daron, 1970) and facultative (Chan *et al.*, 1973) strains of Bacillus stearothermophilus and extreme thermophiles (Ray *et al.*, 1971), raising the growth temperature produced a shift in the fatty acid composition in the membrane lipid. The shift was to fatty acids which, in general, had higher melting points. These results established a definite correlation between the fatty acid content of the membrane and the temperature of growth, and are consistent with the theory that the ability of an organism to grow at an elevated temperature is related to the melting temperature of its membrane lipids. It is tempting to speculate that a specific fatty acid composition of the membrane may set the maximum growth temperature.

The studies demonstrating that the fatty acid composition of the membrane lipid affects the temperature characteristics of a variety of biological processes lends support to this hypothesis.

In our laboratory we have shown that the membrane content of the cell and the protein-to-lipid ratio of the membrane of an obligate thermophile increases as the growth temperature is increased (Table 4). As the growth temperature is increased, the protein content of the membrane increases and the lipid content decreases. Although a detailed analysis of the membrane lipid of this organism has not been done, it is not unreasonable to assume that, as the growth temperature is increased, there is a shift to fatty acids having higher melting points. Oo and Lee (1971) reported similar but larger changes in the membranes of a facultative thermophile grown at 37°C and at 55°C (values in parentheses in Table 4).

The increase in the amount of protein in the membrane as the growth temperature is raised may reflect changes in

237

Table 4

Effect of Growth Temperature on Membrane Content and Composition
of B. stearothermophilus[a]

Growth temperature (°C)	Membrane (%)	Protein (µg/mg membrane)	Lipid	Ratio of protein to lipid
37	-- (20.5)[b]	-- (646)[b]	-- (213)[b]	-- (3.03)[b]
55	16.5 (16.4)[b]	650.0 (803)[b]	178.2 (106)[b]	3.65 (7.58)[b]
60	16.9	664.1	155.5	4.27
65	17.8	680.0	130.3	5.22
70	18.4	690.1	118.2	5.85

[a]From Bodman and Welker (1969) and Wisdom and Welker (1973).

[b]Numbers in parentheses from Oo and Lee (1971).

metabolic or biosynthetic pathways of these organisms. Jung and co-workers (1974) reported that a facultative thermophile grown at 37°C and at 55°C in a complex medium were identical in maximum specific growth rates and yields in cell mass. Cellular extracts of the bacterium, however, showed remarkable differences in the activity levels of two enzymes. Relatively high activities of alcohol dehydrogenase and glyceraldehyde-3-phosphate dehydrogenase were found in the extracts of the 55°C cultures.

Extracts of cells grown at 37°C contained no detectable alcohol dehydrogenase activity and relatively low levels of glyceraldehyde-3-phosphate dehydrogenase activity. They concluded that during exponential growth at 55°C the organism obtains its energy from aerobic respiration and fermentation, whereas during exponential growth at 37°C energy is obtained mainly by aerobic respiration. It is clear from these studies that a change in the protein content of membrane can be a result of a shift in metabolic pathways. Changes in the membrane lipid composition in response to increasing the growth temperature can also be viewed as temperature effects on the enzymes of the various biosynthetic pathways.

In what follows I will describe some experiments in which we attempted to demonstrate a direct correlation between growth temperature, changes in the protein and lipid composition of the membrane, and thermostability of membrane-bound enzymes.

We reported that protoplasts of an obligate thermophile in contrast to protoplasts or spheroplasts of mesophiles are exceptionally resistant to osmotic rupture and are not sensitive to mechanical manipulation. Therefore, they do not need a stabilizing medium other than divalent cations to keep them from rupturing (Bodman and Welker, 1969). The thermostability of these protoplasts is enhanced as the cell growth temperature is raised (Wisdom and Welker, 1973). It seemed reasonable that these protoplasts with their unique stability could be used to study the relationship between membrane structure and heat resistance of thermophilic organisms. Specifically they can be used in place of the intact cell to study differences in the temperature sensitivity of biological molecules *in vivo* and *in vitro*. In addition, protoplasts prepared from cells grown at various temperatures can be used to examine the role of the membrane in protecting cellular constituents and membrane-bound enzymes against thermal inactivation.

Protoplast stability is enhanced as the growth temperature of the culture is raised (Table 5). Protoplasts were prepared from cells grown at each of the indicated temperatures, suspended in high Mg^{2+} buffer, and held at various temperatures for one hour. The results are presented as percentage of protoplast rupture. As the growth temperature was increased, the thermostability of the protoplasts was enhanced. Protoplasts suspended in buffer, or EDTA buffer, rupture at all temperatures regardless of the temperature at which the cells were grown. These results indicate that the enhanced thermostability cannot be explained by a divalent cation protection. As the growth temperature is raised, the enhancement of protoplast stability can be correlated with an increase in the membrane content of the cell, and in the protein of the membrane. The lipid content of the membrane, however, decreases. Appropriate controls were run which eliminated the presence of endogenous enzymes that would lyse the protoplast membrane.

Since we have demonstrated that protoplast stability is enhanced as the growth temperature is raised, we were interested in determining whether specific intracellular constituents had also gained an enhanced thermostability. Two enzymes were selected on the basis of their cellular localization. The enzymes were alkaline phosphatase, an internal or a loosely bound enzyme (peripheral protein), and the membrane-bound NADH oxidase of the terminal respiratory system (integral protein). EDTA lysis of protoplasts releases all of the alkaline phosphatase, whereas the NADH oxidase remains associated with the protoplast "ghosts" (Wisdom and Welker, 1973).

Data relating to the correlation between protoplast stability and the thermostability of these enzymes is shown in Table 6. Protoplasts were prepared from cells grown at various temperatures. Each protoplast suspension (55°C-protoplasts, etc.) in Mg^{2+} buffer was divided into 13 portions. Protoplasts in five of the samples were ruptured by the addition of EDTA. The membranes were collected by centrifugation, the supernatant fluids were adjusted to an equal protein concentration, and designated "free alkaline phosphatase." Five of the remaining samples were designated "internal alkaline phosphatase," and three were designated as "NADH oxidase." Samples of the "free" and "internal" alkaline phosphatase were incubated for 1 hour at 6°C (control), 55°, 60°, 65°, or 70°C. After incubation, "internal alkaline phosphatase" was released by the addition of EDTA.

Table 5

Effect of Growth Temperature on Protoplast Rupture, and the Membrane Composition and Content of *Bacillus stearothermophilus* 1503-4R[a]

Growth temperature	Percentage of protoplast rupture[b]						Membrane[c]	Protein[d]	Lipid[d]	Ratio of protein to lipid
	55°C	60°C	65°C	70°C	75°C	80°C				
55°C	0	0	90	100	100	100	16.5	650.0	178.2	3.65
60°C	0	0	0	80	100	100	16.9	664.1	155.5	4.27
65°C	0	0	0	0	70	100	17.8	680.0	130.3	5.22
70°C	0	0	0	0	10	100	18.4	690.1	118.2	5.84

[a] From Bodman and Welker (1969) and Wisdom and Welker (1973).

[b] Protoplast rupture was quantitated by direct counts with a phase-contrast microscope, and by the liberation of 280 and 260 nm absorbing material.

[c] The dry weight of the membranes was determined gravimetrically, and the results are expressed as the percentage of the dry cell weight.

[d] Values expressed as micrograms per milligram membrane.

Table 6

Thermostability of Free and Internal Alkaline Phosphatase and NADH Oxidase[a]

Growth temperature	Percentage of inactivation of alkaline phosphatase								Percentage of inactivation of NADH oxidase		
	Free				Internal						
	55°C	60°C	65°C	70°C	55°C	60°C	65°C	70°C	75°C	80°C	85°C
55°C	11	15	100	100	0	0	87	100	100	100	100
60°C	·12	14	100	100	0	0	0	92	90	100	100
65°C	11	15	100	100	0	0	0	20	75	84	100
70°C	12	16	100	100	0	0	0	5	0	10	87

[a]From Wisdom and Welker (1973).

The "NADH oxidase" samples were incubated for 1 hour at 75°, 80° and 85°C, followed by the addition of EDTA. Enzyme activity in each of the samples was measured and the percentage of inactivation was calculated.

NADH oxidase was less sensitive to thermal inactivation as the growth temperature was increased. The same results were obtained when this experiment was run with protoplast "ghosts." The apparent thermostability of NADH oxidase is not dependent on the intactness of the protoplast or the presence of divalent cations.

Free alkaline phosphatase was inactivated to the same extent regardless of the growth temperature. The extent of the inactivation was the same in the presence of excess Mg^{2+}.

There is a good correlation between the extent of protoplast rupture and the thermal inactivation of alkaline phosphatase. The data in Table 7 were taken from the two previous tables and demonstrate that alkaline phosphatase is not inactivated in the intact protoplast. Once released from the protoplast, however, alkaline phosphatase is inactivated at 65°C and 70°C (Table 6). Some inactivation of alkaline

Table 7

Effect of Growth Temperature on Protoplast Rupture and the Thermostability of Internal Alkaline Phosphatase[a]

Growth temperature	Percentage of protoplast rupture		Percentage of inactivation of alkaline phosphatase	
	65°C	70°C	65°C	70°C
55°C	90	100	87	100
60°C	0	80	0	92
65°C	0	0	0	20
70°C	0	0	0	5

[a]From Tables 5 and 6.

phosphatase occurs in intact 65°C- and 70°C-protoplasts held at 70°C.

These results indicate that, in the intact protoplast, alkaline phosphatase is protected from thermal inactivation. The apparent thermostability of the NADH oxidase system increases as the growth temperature is raised. It is difficult, however, to differentiate between inherent protein stability and a protective association of this enzyme system with the protoplast membrane.

It is possible to functionally reconstitute NADH oxidase in aggregated membrane after dialysis against Mg^{2+} buffer of nonionic detergent solubilized membranes (Welker, unpublished data). It would then be possible to mix solubilized membranes prepared from cells grown at 55°C (55°C-membranes) or 70°C (70°C-membranes) in varying proportions, and then determine the thermostability of NADH oxidase in the various aggregated membrane preparations.

The 55°C-membranes and 70°C-membranes were differentiated by labeling each membrane preparation with a different isotope. The data for 55°C-membranes (Table 8) show that labeled acetate is incorporated only into the lipid portion of the membrane. The specific activity of the 70°C-membrane lipid was similar to that of the 55°C-membrane lipid.

The recovery of protein, lipid, and reconstituted NADH oxidase in aggregated 55°C-membrane material is shown in Table 9. The protein-to-lipid ratio in aggregated membrane is similar to that of the intact protoplast membrane. The effectiveness of the aggregation procedure is demonstrated by the relatively high recovery of protein and lipid. The extent of NADH oxidase recovery varies from 50-70%. The incomplete recovery of NADH oxidase activity can be a result of an incorrect association in the membrane aggregates or to the inactivation of the enzyme complex during the detergent solubilization. Essentially identical recoveries were obtained with 70°C-membranes.

The thermostability of reconstituted NADH oxidase in aggregated membrane is shown in Table 10. In these experiments we used membranes prepared from cells grown at 55°C or at 70°C. The 55°C-membranes were labeled with ^{14}C and the 70°C-membranes labeled with ^{3}H. Detergent solubilized 70°C-membranes were held at 90°C for 1 hour to inactivate NADH oxidase. The solubilized membranes were mixed in various

Table 8

Incorporation of Radioactive Acetate into Membrane Lipid[a]

Fraction	Complex Medium Containing[b]	
	[^{14}C] Acetate	[^{3}H] Acetate
I. Protoplast Membrane		
Weight (mg)	2.05	2.47
Total Radioactivity (CPM)	3.47×10^{5}	3.58×10^{5}
II. Membrane Lipid		
Weight (mg)	0.38	0.42
Total Radioactivity (CPM)	3.45×10^{5}	3.56×10^{5}
III. Membrane Protein		
Weight (mg)	1.33	1.59
Total Radioactivity (CPM)	0	0

[a]Unpublished data.

[b]Sodium [methyl – ^{3}H] acetate (100 mCi/mmole) or sodium
[1, 2 – ^{14}C] acetate (32.9 mCi/mmole) were added to TYF medium at
0.1 μCi/ml and 0.5 μCi/ml, respectively.

Table 9

Protein and Lipid Content and Functionally Reconstituted NADH Oxidase in Aggregated Membrane[a]

Fraction[b]	Protein (mg)	Lipid (mg)	$[^{14}C]$ Lipid (CPM)	NADH Oxidase (10^5 units)	Ratio of Protein:Lipid
I. Protoplast Membrane	3.120	0.86	7.80×10^5	20	3.63
II. Aggregated Membrane Suspension					
a. 27,000 × g pellet	2.880	0.83	7.54×10^5	11	3.47
b. 27,000 × g super-natant fluids	0.154	ND[c]	ND	ND	--
Per cent Recovery	92.2%	96.5%	96.7%	55%	--

[a] Unpublished data.

[b] Protoplasts and protoplast membranes were prepared as described by Bodman and Welker (1969). Protoplast membranes were solubilized with Triton X-100.

[c] ND = none detected.

Table 10

Thermostability of Functionally Reconstituted NADH Oxidase
in Aggregated Membrane

Per cent solubilized membrane in mixture		Per cent recovery in aggregated membrane		Per cent inactivation of NADH oxidase		
70°C [^{3}H]^a	55°C [^{14}C]	[^{3}H]-lipid	[^{14}C]-lipid	75°C	80°C	85°C
0	100	0	97.5	100	100	100
25	75	23	74	30	72	100
50	50	46	48	7	22	95
75	25	73	23	4	17	90
100	0	92	0	--	--	--
100^b	0	95	0	5(0)	18(10)	94(87)

^{a}Detergent solubilized 70°C [^{3}H]-membranes were heated at 90°C for 1 hour to inactivate NADH oxidase.

^{b}Detergent Triton X-100 solubilized 70°C [^{3}H]-membrane held at 37°C for 1 hour.

Per cents of NADH oxidase inactivation in nonsolubilized membranes at 75°C, 80°C, and 85°C are in parentheses.

proportions, and the mixtures were dialyzed against Mg^{2+} buffer to effect aggregation. The recovery of each label in the aggregated membrane was determined. The recovery of NADH oxidase in the aggregated membrane varied between 55% and 60%.

The heating procedure did not significantly affect the membrane aggregation (last line; unheated control). However, the NADH oxidase in aggregated membrane was somewhat more sensitive to thermal inactivation when compared to the inactivation of NADH oxidase in the intact protoplast membrane (numbers in parentheses).

In mixtures containing 50% or 75%, 70°C-membrane material, the extent of NADH oxidase inactivation was similar to that observed in the unheated 70°C-membrane control sample (last line). The thermostability of NADH oxidase is still considerably enhanced when the mixture contains only 25% 70°C-membrane material. These results suggest that 70°C-membrane material provides some degree of protection against thermal inactivation of the NADH oxidase present in 55°C-membranes. The recovery of membrane protein in the reaggregated material was not determined.

Arrhenius plots of the NADH oxidase activity in protoplast membranes prepared from cells grown at different temperatures are shown in Figure 3. The Arrhenius plots have

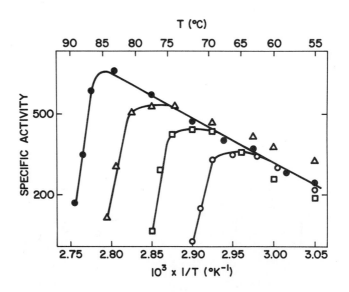

virtually identical slope until the high temperature denaturation occurs.

The transition temperature for the high temperature denaturation of NADH oxidase was dependent for the most part on the growth temperature of the cells from which the membranes were obtained. Inactivation of the membrane-bound NADH oxidase was observed between 64°C and 66°C in the 55°C-membranes; 70°C and 75°C in the 60°C-membranes; 76°C and 79°C in the 65°C-membranes; and 84°C and 86°C in the 70°C-membranes. Temperatures below 55°C were not used, so a low temperature denaturation was not observed (Brandts, 1967).

Preliminary studies using differential scanning calorimetry of 55°C-membranes and extracted lipids showed that a reversible endothermic phase transition occurs in the temperature range of 37°-54°C. Irreversible protein denaturation was observed between 60°C and 75°C. If we assume that above 54°C all the membrane lipids are in the liquid-crystalline state, the inactivation of NADH oxidase must be due to protein denaturation rather than to a lipid phase transition.

Arrhenius plots of the NADH oxidase activity in reaggregated membranes are shown in Figure 4. Only the reaggregated 55°C- and 70°C-membrane samples and a reaggregated sample containing 75%, 70°C-membrane (NADH oxidase inactivated by heating at 90°C for 1 hour) and 25%, 55°C-membrane are shown. The three Arrhenius plots have virtually identical slopes with no change in the activation energy until the high temperature denaturation occurs. As shown in Table 11, the temperature at which NADH oxidase is inactivated is slightly lower in the reaggregated membrane than in the intact membrane. Inactivation of the NADH oxidase in the reaggregated membrane mixture was observed between 73°C and 74°C. These results provide additional evidence that 70°C-membrane protects the 55°C-membrane NADH oxidase from thermal inactivation.

Fig. 3. Arrhenius plots of NADH oxidase activity in 55°C-membranes (o), 60°C-membranes (□), 65°C-membranes (△), and 70°C-membranes (●).

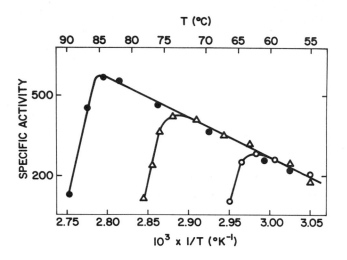

Fig. 4.   *Arrhenius plots of NADH oxidase activity in reaggre-*
gated 55°C-membranes (o), reaggregated 70°C-mem-
branes (●), and reaggregated membrane consisting
of 75%, 70°C-membrane (NADH oxidase was inactivated
by heating at 90°C for 1 hour) and 25%, 55°C-mem-
brane (△).

We conclude from these studies that the thermostability
of alkaline phosphatase of this organism was not affected by
the growth temperature. This enzyme is protected from thermal inactivation in the intact protoplast. The extent of
this protection appears to be correlated with the composition
of the protoplast membranes. It is possible that the membrane acts as an insulator to the transfer of heat from the
environment so that alkaline phosphatase and other soluble
enzymes are protected against thermal denaturation. We have
some evidence that some lipid is associated with the alkaline phosphatase. This finding, if verified, would suggest
that this enzyme is attached to the membrane *in vivo* (peripheral enzyme).

Table 11

Temperature of Inactivation of NADH Oxidase
in Intact and Reaggregated 55°C- and 70°C-membranes[a]

Membrane	Temperature range of inactivation of NADH oxidase
55°C-membranes	64°-66°C
Reaggregated 55°C-membranes	61°-63°C
70°C-membranes	84°-86°C
Reaggregated 70°C-membranes	83°-85°C

[a]*From Figures 3 and 4.*

The thermostability of the membrane-bound NADH oxidase system also increases as the temperature of growth is raised, but is not dependent on an intact protoplast. It is still difficult, however, to differentiate between inherent protein stability and a protective association of this enzyme system with the protoplast membrane.

From the preliminary data presented here, we can conclude that there is a correlation between the temperature of growth and the structure of the membrane. The effect of growth temperature is probably to modify the activity and/or biosynthesis of enzymes concerned with maintaining the structural and functional integrity of the membrane. As a result of these varied changes in membrane structure the thermal stability of some cytoplasmic and membrane components is enhanced.

There is a paucity of data relating to the effect of temperature on membrane proteins. The data presented here are of a preliminary nature and a good deal more must be done to determine the chemical and physical changes occurring in the membrane, as well as the regulatory mechanisms that bring about these changes.

Membrane reconstitution appears to offer the most

promising approach to the investigation of this very interesting problem. Reconstitution studies should ideally include: (1) the solubilization of the biomembrane into individual components (phospholipid, protein, etc.), (2) the chemical, biochemical, and biophysical characterization of the solubilization products (e.g., the stabilization of membrane-bound enzymes would be necessary in order to assess their thermostability), and (3) their reassembly into functional aggregates (homogeneous and heterogeneous) which ultimately should be identical to the native membranes in both structure and function.

Data obtained from reconstitution experiments along with the information obtained from investigations with purified cellular components should contribute to our understanding of the molecular basis of thermophily.

REFERENCES

AMELUNXEN, R., and LINS, M. (1968). Arch. Biochem. Biophys. 125, 765.

BODMAN, H., and WELKER, N. E. (1969). J. Bacteriol. 97, 924.

BOYER, P. D., LUM, F. G., BALLOU, G. A., LUCK, J. M., and RICE, R. G. (1946). J. Biol. Chem. 162, 181.

BOYER, P. D., BALLOU, G. A., and LUCK, J. M. (1947). J. Biol. Chem. 167, 407.

BRANDTS, J. F. (1967). In "Thermobiology," (A. H. Rose, Ed.), p. 25. Academic Press, New York.

BUBELA, B., and HOLDSWORTH, E. S. (1966). Biochim. Biophys. Acta 123, 376.

BURTON, K. (1951). Biochem. J. 48, 458.

CHAN, M., HIMES, R. H., and AKAGI, J. M. (1971). J. Bacteriol. 106, 876.

CHAN, M., VIRMANI, Y. P., HIMES, R. H., and AKAGI, J. M. (1973). J. Bacteriol. 113, 322.

CHO, K. Y., and SALTON, M. R. J. (1964). Biochim. Biophys. Acta 84, 773.

DARON, H. H. (1970). J. Bacteriol. 101, 145.

DeKRUYFF, B., DEMEL, R. A., and VAN DEENEN, L. L. M. (1972). Biochim. Biophys. Acta 255, 331.

DeKRUYFF, B., VAN DIJCK, P. W. M., GOLDBACH, R. W., DEMEL, R. A., and VAN DEENEN, L. L. M. (1973). Biochim. Biophys. Acta 330, 269.

EIDLIC, L., and NEIDHARDT, F. C. (1965). J. Bacteriol. 89, 706.

ENGELMAN, D. M. (1970). J. Mol. Biol. 47, 115.

INGRAHAM, J. L. (1962). In "The Bacteria," (I. C. Gunsalus and R. Y. Stanier, Eds.), p. 265. Academic Press, New York.

JUNG, L., JOST, R., STOLL, E., and ZUBER, H. (1974). Arch. Microbiol. 95, 125.

KIMELBERG, H. K., and PAPAHADJOPOULOS, D. (1972). Biochim. Biophys. Acta 282, 277.

MACHTIGER, N. A., and FOX, C. F. (1973). Ann. Rev. Biochem. 42, 575.

McELHANEY, R. N., and TOURTELLOTTE, M. E. (1970). Biochim. Biophys. Acta 202, 120.

McELHANEY, R. N. (1974). J. Mol. Biol. 84, 145.

MELCHIOR, D. L., MOROWITZ, H. L., STURTEVANT, J. M., and TSONG, T. Y. (1970). Biochim. Biophys. Acta 219, 114.

OO, K. C., and LEE, K. L. (1971). J. Gen. Microbiol. 69, 287.

OVERATH, P., SCHAIRER, H. C., and STOFFEL, W. (1970). Proc. Nat. Acad. Sci. U.S.A. 67, 606.

RAY, P. H., WHITE, D. C., and BROCK, T. D. (1971). J. Bacteriol. 106, 25.

SHEN, P. Y., COLES, E., FOOTE, J. L., and STENESH, J. (1970). J. Bacteriol. 103, 479.

SINGER, S. J. (1971). In "Structure and Function of Biological Membranes," (L. I. Rothfield, Ed.), p. 145. Academic Press, New York.

SINGER, S. J., and NICOLSON, G. L. (1972). Science 175, 720.

SINGLETON, Jr., R., and AMELUNXEN, R. E. (1973). Bacteriol. Rev. 37, 320.

STEIM, J. M. (1972). In Proc. 8th FEBS Meeting, Amsterdam, 28, 185. North-Holland Publishing Co., Amsterdam.

STEIM, J. M., TOURTELLOTTE, M. E., REINERT, J. C., McELHANEY, R. N., and RADER, R. L. (1969). Proc. Nat. Acad. Sci. U.S.A. 63, 104.

WILSON, G., and FOX, C. F. (1971). J. Mol. Biol. 55, 49.

WISDOM, C., and WELKER, N. E. (1973). J. Bacteriol. 114, 1336.

THE BIOLOGICAL SIGNIFICANCE OF ALTERATIONS IN THE FATTY ACID COMPOSITION OF MICROBIAL MEMBRANE LIPIDS IN RESPONSE TO CHANGES IN ENVIRONMENTAL TEMPERATURE

Ronald N. McElhaney

INTRODUCTION

It has been known for many years that the fatty acid composition of the cellular lipids of many microorganisms (Terroine et al., 1927; Terroine et al., 1930; Pearson and Raper, 1927; Gaughran, 1947a and b; Kates and Baxter, 1962; Marr and Ingraham, 1962; Kates and Hagen, 1964; Daron, 1970; Ray et al., 1971a and b; Singleton and Amelunxen, 1973; McElhaney, 1974a; Souza et al., 1974), as well as of insects (Fraenkel and Hopf, 1940) and the higher plants and animals (Leathes and Raper, 1925; Bĕlehrádek, 1931; Hilditch, 1956; Deuel, 1955), changes rapidly, and often markedly, in response to alterations in the environmental temperature. These alterations in lipid composition are manifested most commonly as changes in the relative proportions of the various fatty acid classes present, less commonly as a change in the average chain length of the fatty acyl groups, or both types of change may occur simultaneously. An increase in the environmental temperature normally results in the production of lipids containing a relatively lower proportion of straight-chain saturated fatty acids. In addition, increasing the environmental temperature often (but not always) results in the production of membrane lipids having a greater average chain length. Although historically these

characteristic changes in fatty acid composition in response
to shifts in the environmental temperature were described
only in mesophilic organisms, several studies have demonstrated that psychrophilic (Kates and Baxter, 1962; Kates and
Hagen, 1964) and thermophilic (Daron, 1970; Ray *et al.*, 1971a;
Ray *et al.*, 1971b; Souza *et al.*, 1974) microorganisms also
exhibit similar responses to changes in growth temperature.

Several studies of the lipid compositions of psychrophilic, mesophilic, and thermophilic microorganisms belonging
to the same genus reveal that a relationship also exists
between the fatty acid composition of the membrane lipids and
the growth temperature range characteristic of a particular
organism (Kates and Baxter, 1962; Kates and Hagen, 1964;
Shen *et al.*, 1970; Chan *et al.*, 1971). The general nature of
these comparative differences in fatty acid composition,
which presumably represent long-term evolutionary adaptations
to different thermal environments, is qualitatively similar
to the short-term response of individual microorganisms to
shifts in growth temperature. That is, psychrophilic microorganisms are typically characterized by relatively high proportions of unsaturated or anteisobranched fatty acyl groups
in comparison to their mesophilic counterparts, whereas the
closely related thermophilic species contain high proportions
of straight-chain saturated and isobranched fatty acids. In
addition, the average length of the hydrocarbon chains tends
to increase as one proceeds from the psychrophilic members of
a genus toward the thermophilic species. Thus, the lipids
from organisms which have evolved to grow at lower temperatures (or which are growing near the lower boundary of their
growth temperature range) consistently exhibit a relative
enrichment in fatty acids characterized by having relatively
low melting points, whereas those adapted to growth at high
temperatures (or which are growing near the upper limit of
their temperature range) are enriched in the higher-melting
fatty acid species. This positive relationship between the
environmental temperature and the melting points of the
cellular lipids led Heilbrunn (1924) and Bělehrádek (1931) to
postulate that the melting of the cellular lipids might set
an upper limit for cellular growth. Brock (1967) later suggested that the thermostability of the cell membrane, perhaps
as influenced either directly or indirectly by the fatty acid
composition of its lipid constituents, may be a major factor
in determining the growth temperature range of microorganisms.
In this paper I will attempt to reformulate Brock's hypothesis
in a more precise and explicit manner in light of recent
advances in our understanding of the structure and function

of biological membranes and the important role of membrane lipids therein. I will also present evidence that I believe strongly supports the notion that adaptive alteration in the fatty acid composition of the membrane lipids represents a major mechanism by which microbial systems ensure their survival not only at very low or very high temperatures, but also over the widest possible temperature range.

It has now been firmly established that the classical lipid bilayer, schematically depicted in Figure 1, is a central structural feature of all biological membranes thus far

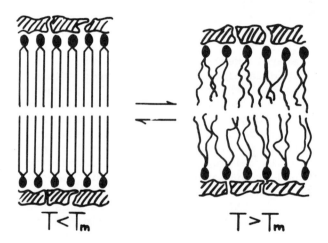

$$T < T_m \qquad\qquad T > T_m$$

Fig. 1. A schematic representation of the gel to liquid-crystalline membrane lipid phase transition. For simplicity, the proteins are shown on the surface of the lipid bilayer, although good evidence exists that some membrane proteins penetrate partially into or traverse the bilayer structure.

examined (Singer and Nicolson, 1972; Bretscher, 1973; Capaldi, 1974; McElhaney, 1974c). A unique feature of phospholipid bilayers which do not contain appreciable quantities of cholesterol or closely related sterols is their ability to undergo a reversible, thermotropic gel to liquid-crystalline phase transition, which arises from a cooperative melting of the hydrocarbon chains in the interior of the bilayer. Unlike phase transitions between the various other liquid-crystalline mesophases, the selective melting of the phospholipid fatty acid chains does not result in any gross molecular rearrangement, so that the general bimolecular leaflet structure exists both above and below the phase transition temperature. However, the properties of the gel and liquid-crystalline states are quite different. In the gel or solid state, the hydrocarbon chains exist predominantly in the all-trans extended conformation and are aligned nearly perpendicularly to the plane of the membrane. The fatty acyl chains can thus undergo a close hexagonal packing, resulting in a minimal cross-sectional area per phospholipid molecule (40-45 $Å^2$) and a near maximum bilayer thickness (50-55 Å, depending on the fatty acid chain length). Both intra- and intermolecular motion is severely restricted in these rigid, relatively impermeable lipid lamellae, and the bilayer structure exhibits a high degree of both short-range and long-range order. In contrast, the liquid-crystalline state is characterized by a fair amount of long-range order combined with a great deal of short-range disorder. The hydrocarbon chains now exist in a partially melted condition and the general bilayer structure is preserved, largely by electrostatic interactions between the polar head groups of the phospholipid molecules and hydrophobic forces. Near the carbonyl group, the hydrocarbon chain is only slightly disordered but becomes nearly "liquid" near the methyl terminal end of the chain. A gradient of increasing fluidity thus exists as we proceed toward the core of the lipid bilayer, but on the average, the hydrocarbon chain exists in a condition midway between the liquid and solid state. Due to the greatly increased areas now occupied by the fatty acyl groups, and the presence of numerous gauche conformations in the hydrocarbon chain, the cross-sectional areas of the phospholipid molecules are now much greater (60-70 $Å^2$) and the bilayer is considerably thinner (40-45 Å). The liquid-crystalline bilayer is thus a fairly loosely packed, fluid, and relatively permeable structure characterized by a substantial amount of intra- and intermolecular motion (Singer and Nicolson, 1972; Bretscher, 1973; Capaldi, 1974; McElhaney, 1974c). It should be noted that in biological membranes, the gel to liquid-crystalline phase transition is not sharp (does not occur at a single discrete temperature) due to heterogeneity in the structure of the polar head groups and fatty acyl chains. Thus the phase transition is broad and really consists of a progressive lateral phase separation wherein domains of gel and liquid-crystalline phases exist simultaneously.

Since the gel to liquid-crystalline phase transition is essentially a hydrocarbon chain melting phenomenon, the temperature at which the transition occurs will be markedly dependent on the nature of the fatty acids esterified to the membrane lipids. As the strength of the van der Waal's attractive forces between adjacent lipid hydrocarbon chains exhibits a nearly direct dependence on the number of interacting methylene groups and a sharp inverse dependence on the distance between these groups, the transition temperature would be expected to vary considerably with the chain length and geometry of the fatty acyl chains. This is illustrated in Table 1, where the melting points and gel to liquid-crystalline phase transition temperatures of a number of fatty acids commonly found in microbial membrane systems are presented. We can see that the straight-chain saturated fatty acids, which exist essentially as linear rods at low temperatures, have as a class the highest melting points. The melting points of the saturated fatty acids decrease as the length of the hydrocarbon chain decreases. A methyl branch introduced at the end of a hydrocarbon chain interferes only minimally with close packing, so the isobranched fatty acids exhibit melting points only slightly below their unbranched analogues. A methyl branch introduced at the antepenultimate carbon atom is much more disruptive, resulting in considerably reduced melting temperatures for the anteisobranched fatty acids. The kink produced by the introduction of a cis-double bond near the center of the hydrocarbon chain produces maximum disruption of close packing so that the monounsaturated fatty acids are the lowest-melting of the series. Table 1 illustrates some other important points. The melting point of the pure, anhydrous free fatty acids is only a fair qualitative predictor of the biologically relevant parameter which is the phase transition temperature of the mixed, fully hydrated fatty acid esterified in a phospholipid or glycolipid molecule. The phase transition temperatures of the pure diacylphosphatides in water are lower than for the free fatty acids, and the transition temperature is considerably more sensitive to the chain length and geometry of the hydrocarbon chains. We have some preliminary evidence with cyclopropane fatty acids that the melting point

Table 1

The Melting Points and Gel to Liquid-crystalline Transition
Temperatures of Some Fatty Acids
Commonly Found in Microbial Membranes

Fatty acid[a]	Melting point (°C)	Transition temperature (Diacyl phosphatidylcholines) (°C)
18:0	70.1	60
16:0	62.9	41
14:0	54.4	23
$18{:}0^{i}$ [b]	69.5	?
$16{:}0^{i}$ [b]	62.4	?
$17{:}0^{a}$ [b]	36.8	?
$15{:}0^{a}$ [b]	23.0	?
$18{:}1_{c}$ [c]	13.4	-22
$16{:}1_{c}$ [c]	~0.0	?

[a]*Fatty acids are designated by the number of carbon atoms followed by the number of double bonds, if any, present in the molecule.*

[b]*The superscripts i and a indicate a methyl group attached to the penultimate carbon atom (an isobranched acid) and antepenultimate carbon atom (an anteisobranched acid), respectively.*

[c]*The subscript c denotes the cis configuration.*

of the pure acids may not even be a good qualitative predictor of transition temperature (unpublished data). In
addition, Table 1 demonstrates that the two types of branched-
chain fatty acids commonly found in certain bacteria should
not be lumped together. The common isobranched fatty acids
doubtlessly have transition temperatures below, but rather
close to the saturated species of similar carbon number,
whereas the anteisobranched acids more closely resemble the

unsaturated fatty acids. Indeed, as mentioned earlier, these two classes of fatty acids often behave quite differently in response to changes in environmental temperature (Daron, 1970; Ray et al., 1971a; Souza et al., 1974; Shen et al., 1970). Finally, the gaps in this table point up the need for more data on the phase behavior of pure natural and synthetic phospholipids of known structure.

Considering the fact that all microbial membranes studied to date are known to contain extensive areas of lipid bilayer, which may exist in either the gel or liquid-crystalline state depending on the environmental temperature, we can now speculate as to what properties of the lipid bilayer structure itself might set lower and upper temperature limits for cell growth, ignoring for a moment nonmembrane proteins and other cellular constituents. It is quite possible that the minimum growth temperature could be directly determined by the lower boundary of the gel to liquid-crystalline phase transition. That is, the microbial membrane may cease to function when all, or a large part, of the membrane lipid becomes solid. Indeed, some theoretical (Luzzati and Husson, 1962; Chapman, 1967; Reiss-Husson and Luzzati, 1967) and experimental (Steim et al., 1969; Overath et al., 1970) evidence exists which suggests that the rigid, tightly packed gel state will not support normal membrane function. The possible effects of the physical state of the membrane lipids on the upper temperature limit of growth are less clear-cut. There is no theoretical basis for supposing that the maximum growth temperature would be directly determined by the upper boundary of the membrane lipid phase transition, and in fact, the lipids of many microorganisms exist entirely in the liquid-crystalline state even at their optimum growth temperatures (Singer and Nicolson, 1972; Bretscher, 1973; Capaldi, 1974; McElhaney, 1974c). However, the physical state of the membrane lipids could influence the maximum growth temperature in at least three ways. At temperatures above the phase transition, the thermal motion of the fatty acyl chains is known to increase markedly with increasing temperature (Fox and Keith, 1972). Since the hydrophobic core of the lipid bilayer appears to be the major cellular permeability barrier (McElhaney et al., 1973), it is not surprising that the rates at which many ions and nonelectrolytes passively diffuse across cell membranes also increase very rapidly with temperature (McElhaney et al., 1973; Stein, 1967; Diamond and Wright, 1969; De Gier et al., 1971). At high temperatures, the cell membrane may become sufficiently "leaky" so that an adequate intracellular concentration of

261

low molecular weight metabolites can no longer be maintained, even though the cellular permeability barrier remains intact. At higher temperatures, the molecular motion of the hydro- carbon chains may become so great that the lipid bilayer structure itself becomes unstable, resulting in at least transient breakdowns in the permeability barrier, and thus a loss of intracellular macromolecules as well as small mole- cular weight substances. Alternatively, the physical state of the membrane lipids could also determine the maximum growth temperature in an indirect fashion by influencing the conformation of one or more membrane proteins. A great deal of evidence has recently been amassed that many integral membrane proteins require lipid which exists in a particular physical state for proper enzymatic or transport function (Singer and Nicolson, 1972; Bretscher, 1973; Capaldi, 1974; McElhaney, 1974c), or for protection from thermal denatura- tion (Wisdom and Welker, 1973). It may be that membrane lipids which exist in an excessively fluid state are no long- er capable of stabilizing certain membrane-bound proteins in a fully functional state.

If we can accept the postulate that, at least under cer- tain circumstances, the minimum growth temperature of a microorganism may be directly determined by the lower boun- dary of the membrane lipid phase transition, and that the maximum growth temperature may also depend upon the physical state of the membrane lipid, the biological significance of the temperature-induced alterations in the fatty acid compo- sition of the membrane lipids discussed earlier becomes clear. The increase in the relative proportions of the longer-chain saturated and isobranched fatty acids and the concomitant reduction in the levels of the unsaturated and anteisobranched acids normally observed at higher tempera- tures would tend to raise the temperatures of the membrane lipid phase transition boundaries, thereby decreasing the membrane lipid fluidity at high temperatures relative to the conditions which would exist in the absence of such a fatty acid compositional shift. Conversely, the relative increase in the proportion of the lower-melting fatty acids noted at lower temperatures would tend to decrease the phase transi- tion temperature, resulting in an increased fluidity, again relative to a state of fatty acid compositional constancy. These characteristic alterations in membrane lipid fatty acid composition in response to changes in the environmental temperature thus provide a mechanism for minimizing the effect of temperature variations on the physical state of the membrane lipids.[1] The existence of such a homeostatic mechanism for regulating membrane lipid fluidity could provide at least two potentially important evolutionary advantages to microbial systems which developed it. In terms of relatively short-term adaptation to a variety of thermal environments, this compensatory shift in the nature of the fatty acyl groups would permit cellular growth to occur over a wider range of temperatures than would otherwise be possible. Operating over the long term, such a regulatory mechanism would facilitate the evolutionary development of microorganisms with membrane lipid compositions capable of supporting normal membrane function in either extremely hot or extremely cold environments.

With this introduction, I would now like to describe in some detail some recent work from my own and other laboratories which I believe supports the idea that the physical state of the membrane lipids can be important in determining the growth temperature range of microorganisms. This work provides indirect support for the concept that temperature-induced changes in membrane lipid fatty acid composition represent an important mechanism in the adaptation of microbes to extreme thermal environments.

MATERIALS AND METHODS

The organism we utilized in our study of the relationship between the physical state of the membrane lipids, temperature, and cell growth was Acholeplasma laidlawii B

---

[1]*Strictly speaking, the hypothesis as formulated applies only to procaryotic microbes since these organisms do not normally contain appreciable amounts of cholesterol or other steroids in their cell membranes. Since cholesterol is known to broaden or abolish the highly cooperative nature of the gel to liquid-crystalline phase transition, one can not properly speak of discrete phase boundaries in the membrane of most eucaryotic microorganisms. Nevertheless, temperature-induced changes in the fatty acid composition of eucaryotic species would have a similar general effect on membrane lipid fluidity to that observed in procaryotic systems. Thus, the hypothesis in its broad sense also applies to sterol-containing eucaryotic protists.*

(formerly Mycoplasma laidlawii B). This mesophilic microorganism is a member of the class Mollicutes, a diverse group
of primitive procaryotes related to, but distinct from the
bacteria (Smith, 1971). The general simplicity of
A. laidlawii, as well as the absence of a cell wall and
internal membrane systems, makes this organism especially
suitable for investigations of membrane structure and function (McElhaney, 1974c). The small genome of this organism
is reflected in its limited ability to synthesize or cata-
bolize fatty acids. A. laidlawii can biosynthesize only
saturated fatty acids containing 12-18 carbon atoms (Pollack
and Tourtellotte, 1967). Unsaturated or branched-chain fatty
acids are not synthesized, nor is any class of fatty acids
oxidized or otherwise degraded by this organism (Pollack and
Tourtellotte, 1967; McElhaney and Tourtellote, 1969). However, a wide variety of exogenous fatty acids can be incorporated in substantial amounts if present in the growth
medium (McElhaney and Tourtellotte, 1969). The fatty acid
composition of the membrane lipids can thus be dramatically
but controllably altered without affecting the normal quantitative distribution of the membrane lipid classes
(McElhaney et al., 1973) or membrane proteins (Pisetsky and
Terry, 1972). This organism is thus an excellent model system for studies of the effect of variations in the nature of
the hydrophobic core of a biomembrane on various membrane
functions (McElhaney, 1974c). In the study described in
detail here, the effect of changes in membrane lipid physical
state, as influenced by fatty acid composition and temperature on the range of temperatures within which A. laidlawii
can grow, and on the rate of growth within the permissible
ranges, was investigated. The experimental design was simple. A. laidlawii cells were grown in the presence of a number of different fatty acids, selected so as to produce a
wide range of membrane lipid fluidities, and the minimum,
optimum, and maximum growth temperatures as well as growth
rates were determined for each culture. The plasma membranes
from cells grown in the different fatty acids were then isolated, and the lipid phase transition boundaries were determined by differential thermal analysis (DTA). Attempts were
then made to correlate the various growth parameters with the
phase state of the membrane lipids. For complete experimental details, the reader is referred to several recent publi-
cations (McElhaney, 1974a and b).

RESULTS

Relation between Temperature and Fatty Acid Composition

As mentioned earlier, the fatty acids incorporated into the membrane polar lipids of A. laidlawii B can be markedly altered by growth in the presence of appropriate exogenous fatty acids. This is demonstrated in Table 2, where the

Table 2

Fatty Acid Composition of the Total Membrane Lipids
of Acholeplasma laidlawii B
Grown with Various Fatty Acids at 35°C

Fatty acids found	Fatty acid added to the growth medium							
	None	$18:0^a$	16:0	$18:0^i$	$18:1_t$	$17:0^a$	$18:1_c$	$18:2_c$
12:0	5.0	8.7	8.5	4.3	2.9	1.3	2.2	4.3
13:0	3.4	trace	trace	trace	trace	trace	trace	trace
14:0	30.8	12.7	6.9	8.9	4.2	2.6	6.6	6.9
15:0	6.6	trace	trace	trace	trace	trace	trace	trace
16:0	40.1	11.6	80.4	6.8	7.2	2.2	22.1	31.2
17:0	3.2	trace	trace	trace	trace	trace	trace	trace
$17:0^a$	--	--	--	--	--	90.7	--	--
18:0	7.7	58.4	1.9	trace	trace	1.0	4.5	2.6
$18:0^i$	--	--	--	76.4	--	--	--	--
18:1	1.8	4.9	1.6	1.5	85.4	1.5	64.4	3.4
18:2	1.2	3.7	0.6	1.9	0.3	0.7	trace	51.3

[a]*Fatty acids are designated by the number of carbon atoms followed by the number of double bonds, if any, present in the molecule; the subscripts c and t denote the cis or trans configurations, respectively, of these double bonds. The superscripts i and a indicate a methyl group attached to the penultimate carbon atom (an isobranched fatty acid) and the antepenultimate carbon atom (an anteisobranched acid), respectively. The fatty acid composition values are expressed as mole percent.*

fatty acid compositions of the membrane lipids of cells grown
at a single temperature in the presence of various exogenous
fatty acids is presented. In all cases where a saturated,
unsaturated, branched-chain or cyclopropane fatty acid was
supplied in the growth medium, that exogenous acid was incorporated in substantial quantities, forming in fact the bulk
of the total esterified fatty acids. Cells grown without
fatty acid supplementation contained only saturated fatty
acids except for small amounts of oleic and linoleic acid
which were incorporated from residual levels of these unsa-
turates still present in the lipid-poor growth medium
(McElhaney and Tourtellotte, 1969). Palmitic and myristic
acid accounted for over 70% of the total fatty acids present.
Cells grown without exogenous fatty acids most closely resemble palmitate-grown cells in overall fatty acid composition,
except that the average chain length is somewhat less in the
unsupplemented culture.

The effect of altering the environmental temperature on
the fatty acid composition of the total membrane lipid of
this organism was also investigated. The results obtained
when no exogenous fatty acids were supplied are presented in
Table 3. Under these circumstances, of course, nearly all
the esterified fatty acid in the membrane lipids is derived
biosynthetically. It can be seen that the fatty acid composition changes significantly, but not markedly, in response
to changes in environmental temperature. As the temperature
of growth is raised, there is a moderate decrease in the
shorter-chain fatty acids accompanied by a corresponding
increase in the levels of the longer-chain saturates.
Although this result is qualitatively similar to those noted
in other microorganisms (see the Introduction), the magnitude
of the fatty acid composition shift is slight, producing a
shift in the phase transition temperature of only about 3°C
(unpublished results).

The effect of growth temperature on fatty acid composition when A. laidlawii B is grown in the presence of an exogenous fatty acid such as oleate is presented in Table 4.
Under these conditions, both the levels of exogenous fatty
acid incorporated and the quantitative distribution of saturated fatty acid biosynthesized by the organism do not change
significantly, even with pronounced variations in the temperature at which these cells are cultivated. This apparent
lack of a mechanism for effectively regulating the physical
state of the membrane lipids in response to changes in
growth, although unusual in microorganisms, is useful in

266

Table 3

Fatty Acid Composition of the Total Membrane Lipids
of Acholeplasma laidlawii B
Growth Without Fatty Acid Supplementation
at Various Temperatures

Fatty acids found	Temperature at which cells were grown (°C)				
	40	35	30	25	20
12:0	4.8	5.0	5.9	8.2	9.6
13:0	3.4	3.4	4.8	6.2	6.5
14:0	30.2	30.8	30.0	28.4	30.1
15:0	7.7	6.6	8.1	8.1	7.2
16:0	40.8	40.1	39.0	37.5	36.9
17:0	2.1	3.2	2.4	1.9	1.2
18:0	8.0	7.7	7.3	7.1	6.5
18:1	1.4	1.8	1.2	1.5	1.2
18:2	1.5	1.2	1.2	1.0	0.8

Table 4

Fatty Acid Composition of the Total Membrane Lipids
of Acholeplasma laidlawii B
Grown with Oleic Acid at Various Temperatures

Fatty acids found	Temperature at which cells were grown (°C)					
	10	15	20	25	30	35
12:0	4.6	6.1	5.6	1.3	1.8	2.2
13:0	trace	trace	trace	trace	trace	trace
14:0	6.6	5.1	6.5	8.1	7.1	6.6
15:0	trace	trace	trace	trace	trace	trace
16:0	22.0	25.5	21.3	22.7	21.6	22.1
18:0	3.0	2.0	1.8	3.3	4.0	4.5
18:1	62.7	61.1	64.8	63.2	63.7	64.4
18:2	1.0	trace	trace	1.3	1.8	trace

these studies, since it simplifies the interpretation of many experimental results and allows us to explore the consequences of the absence of such a mechanism.

Relation between Fatty Acid Composition and the Physical State of the Membrane Lipids

Temperature-based thermograms of isolated membranes derived from A. laidlawii cells grown in the presence of a variety of different fatty acids are shown in Figure 2. These thermograms depict the gel to liquid-crystalline membrane lipid phase transitions discussed previously. Because of the heterogeneous nature of the fatty acid esters present in each membrane system, these phase transitions occur over a 15°-25°C range in temperature. Within the transition temperature range, both the gel and liquid-crystalline phases exist simultaneously, so that the membrane is heterogeneous with respect to the physical state of its lipids (Oldfield and

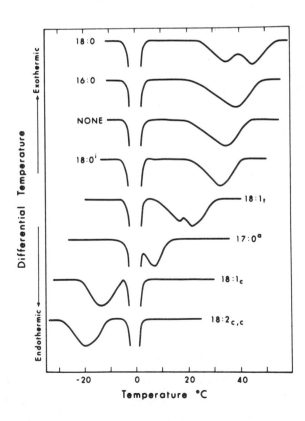

Chapman, 1972). As expected, the transition temperatures are
profoundly influenced by the nature of the fatty acids esterified to the membrane lipids. Membranes enriched in linoleic
acid, for example, undergo the phase transition at temperatures nearly 70°C below those observed for membranes enriched
in stearic acid. The phase transitions were found to be completely reversible and could be abolished by the addition of
cholesterol. Thermograms of dispersions of the total membrane lipid in excess water were nearly identical to those
obtained from isolated membranes of similar fatty acid composition.

*Relation between the Minimum, Optimum, and Maximum Growth
Temperatures and the Physical State of the Membrane Lipids*

The minimum, optimum, and maximum growth temperatures of
A. laidlawii cells grown in the presence of a variety of exogenous fatty acids are presented in Table 5, as well as the
phase transition temperature range and the transition midpoint temperature of the isolated membranes of corresponding
fatty acid composition. It will be noted that the minimum
growth temperature can be most strongly and directly dependent on the physical state of the membrane lipids. The minimum growth temperature of cells grown in the lower-melting
fatty acids does not appear to be determined by the lipid
components of the membrane, since cells supplemented with
linoleic, oleic, or anteisoheptadecanoic acid could not grow
below 8°C. However, the minimum growth temperature of cells

---

*Fig. 2. Temperature-base thermograms of isolated A. laid-
lawii plasma membranes grown without fatty acid supplementation or in the presence of various exogenous
fatty acids. The temperature differentials between
the samples and inert reference material are plotted
as a function of the temperature of the reference,
using a heating rate of 5°C per minute. Cooling the
samples at a rate of 5°C results in essentially
identical curves except that the entire thermogram
is shifted by 2-3°C to lower temperatures. The
large endotherm centered around 0°C is due to melting of the ice from excess water associated with the
membrane preparations.*

Table 5

The Minimum, Optimum, and Maximum Growth Temperatures
and the Membrane Lipid Phase Transition Parameters
of Acholeplasma laidlawii B Cells
Grown in Various Fatty Acids

Fatty acid added	Growth temperatures (°C)			Transition midpoint	Transition range
	Minimum	Optimum	Maximum		
18:0	28	38	44	41	25-55
16:0	22	36	44	38	20-50
None	20	36	44	34	18-45
18:0^i	18	36	44	32	18-42
18:1$_t$	10	36	44	21	5-32
17:0^a	8	36	44	7	0^a-15
18:1$_c$	8	34	40	-13	-22 to -4^a
18:2$_{c,c}$	8	32	38	-19	-30 to -10

^{a}These temperatures are estimates because a portion of
the lipid phase transition endotherms was obscured by the
melting of the ice from the excess water associated with the
membrane preparations.

supplemented with all other fatty acids or grown without
fatty acid always fall above 8°C and between the lower end
of the phase transition range and the transition midpoint,
always being closer to the former. Since the magnitude of
the temperature differential at any given temperature within
the phase transition range is approximately proportional to
the amount of lipid undergoing the phase transition of that
temperature, the progress of the phase transition can be
monitored by analysis of the differential thermal analysis
thermograms. The area above the differential thermal analysis curve between the onset of the transition and a particular temperature can be measured by planimetry and compared to
the peak area of the total transition. The ratio of these

areas should yield an estimate of the proportion of gel to
liquid-crystalline phase that exists at any temperature within the phase transition boundaries. At the minimum growth
temperatures of those cells grown in the higher-melting fatty
acids such an analysis shows that less than one tenth of the
total membrane lipid still exists in the liquid-crystalline
state. As the environmental temperature is lowered to the
point where the hydrocarbon chains of the membrane lipids
approach the fully crystalline state, cell growth ceases.

It is also of interest to note that the broadest range
of growth temperatures is achieved by cells supplemented with
either elaidic or anteisoheptadecanoic acid. Cells supplemented with these two fatty acids are characterized by lipid
phase transitions which occur at intermediate temperatures,
with the lower boundary of the transition positioned below
but fairly close to the absolute minimum growth temperature.

*Relation between Growth Rates and the Physical State
of the Membrane Lipids*

Arrhenius plots of the growth rate, expressed in generations per hour, as a function of temperature are presented
in Figure 3 for cells grown with a variety of exogenous fatty
acids. In all cases, the rate of growth declines with
decreasing temperature, but not in the same manner for cells
enriched with different fatty acids. Arrhenius plots of
three distinct shapes can be noted. Cells grown in oleic or
linoleic acids, and whose lipid phase transition midpoint
temperatures are well below the minimum growth temperature of
these cells, exhibit a simple linear dependence of the
logarithm of the growth rate on reciprocal temperature over
the entire temperature range between the minimum and optimum
growth temperatures. The apparent temperature characteristic
of growth calculated from the slopes of these plots is
16-18 kcal/mole. Cells grown in palmitic or stearic acid,
and whose phase transition midpoint temperatures fall above
the optimum growth temperatures for these cells, also show a
linear relation between the logarithm of the growth rate and
reciprocal temperature in the temperature range between the
minimum and optimum growth temperatures. With palmitate-
and stearate-enriched cells, however, the apparent temperature characteristic of growth is much higher, being approximately 45 kcal/mole. On the other hand, cells grown in anteisoheptadecanoic, elaidic, or isostearic acids, or with no
fatty acid additions, and whose lipid phase transition midpoint temperatures fall between the minimum and optimum
growth temperatures of these cells, show "breaks" or abrupt

271

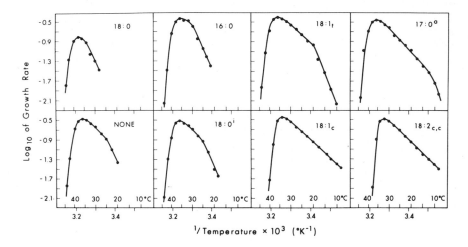

Fig. 3. Arrhenius plots of the relative growth rate of A. laidlawii cells, expressed in generations per hour versus the temperature of growth.

changes in the slopes of their Arrhenius plots at temperatures of about 12°, 20°, 24°, and 25°C, respectively. At temperatures above those at which the break in slope occurs, the temperature coefficient of growth is approximately 16-18 kcal/mole, while at lower temperatures this temperature coefficient rises to values of 40-50 kcal/mole. The temperature at which the break in the Arrhenius plots is noted always falls within the phase transition boundaries, usually being closer to the region of the phase transition midpoint than to either the upper or lower limits of the transition. However, these breaks in the Arrhenius plots do not appear to correlate with one particular and discrete region of the phase transition to the extent observed for the minimum growth temperatures. In general, the temperature of growth does not appear to be altered when up to about half of the membrane lipid is converted from the liquid-crystalline to the gel state, but growth rates decline

272

rapidly with temperature when the membrane lipid begins to
become predominantly crystalline.

As discussed in the previous section, Figure 3 also
illustrates that the minimum, optimum, and maximum growth
temperatures are altered in a characteristic way by variations in the fatty acid composition of the membrane lipids.
However, it is interesting to note that at the optimum
growth temperature, the absolute rates of growth are very
similar for cells grown in the various exogenous fatty acids
or with no fatty acid additions to the growth medium. In all
cases the generation time was approximately three hours. The
only exception was noted for cells grown in stearic acid,
where the maximum growth rate was only half that obtained
with all other fatty acid compositions tested.

*Relation of Maximum Cell Density and the Physical State
of the Membrane Lipids*

The maximum cell densities obtained by cell cultures
grown in the presence of a variety of exogenous fatty acids
or without fatty acid supplementation at their respective
optimum growth temperatures are very similar, thus indicating
that maximum cell density does not depend on the physical
state of the membrane lipids. Stearic-acid-grown cultures
again proved to be exceptional, in that the maximum cell
density obtained by these cells normally varies from 40-50%
of that obtained with other fatty acid supplementations (data
not presented).

DISCUSSION

A definite relation exists between the physical state of
the membrane lipids and the temperature range over which
A. laidlawii cells of various fatty acid compositions can
grow. The absolute minimum growth temperature of 8°C is not
defined by the physical state of the membrane lipid when
cells are enriched in fatty acids giving rise to membrane
lipid phase transitions occurring entirely, or in part, below
this temperature. At temperatures below 8°C, some other
factor, such as the existence of a cold-labile enzyme or
regulatory protein, becomes growth limiting. However, the
growth of A. laidlawii in fatty acids giving rise to lipid
phase transitions occurring above the absolute minimum growth
temperature results in a new, elevated minimum growth temperature, which is clearly determined by the fatty acid composition of the cell membrane. That this organism should not

273

be able to grow at temperatures at which its membrane lipids
exist entirely in the gel state is not surprising, and con-
firms our earlier observation that some of the lipid hydrocarbon chains must be in a liquid-crystalline state to support proper membrane function (Steim *et al.*, 1969). The optimum and maximum growth temperatures also proved to be influenced to a significant degree by the physical state of the
membrane lipids, shifting to lower temperatures when cells
are grown in the presence of the lower-melting fatty acids.
If thermal denaturation of some essential cell protein (or
proteins) defines the optimum and maximum growth temperature,
this observation suggests that this key protein is an intrinsic membrane protein, since its conformation would appear to
be intimately related to the physical state of the membrane
lipids. Alternatively, the decrease in the optimum and maximum growth temperatures noted with the lower-melting fatty
acids could be due to the development of excessive leakiness
or to an actual breakdown in the permeability barrier of
these more fluid membranes at higher temperatures. In earlier studies of the effect of variations in fatty acid composition on the cellular permeability to nonelectrolytes, however, no evidence of a breakdown in the permeability barrier
of A. laidlawii B cells cultured in oleic or linoleic acids
was noted, even at temperatures well above the maximum growth
temperatures noted in this study (McElhaney *et al.*, 1973).
Also, it seems unlikely that cells containing a variety of
different higher-melting fatty acids would all become excessively leaky at the same temperature. I therefore favor the
idea that the effect of the fatty acid composition on the
optimum and maximum growth temperatures is due to their
effect on the conformational stability of one or more membrane proteins. However, more work is obviously required to
clearly differentiate between these and other alternative
explanations.

    In addition to defining the temperature range over which
growth can occur, the physical state of the membrane lipids
can also markedly affect the temperature coefficient of
growth within this temperature range. At temperatures below
the optimum growth temperature, cells which contain up to
about half of their membrane lipids in the gel state are
able to function quite as well as cells whose membrane lipid
exists entirely in the liquid-crystalline state within the
same temperature range, as judged by both their similar absolute growth rates and similar temperature coefficients of
growth of 16-18 kcal/mole. Thus, a heterogeneous state of
lipids *per se* seems not to be detrimental to membrane

function in this organism, provided that the proportion of
solidified lipid does not become too high. However, if the
environmental temperature is lowered, such that more than
about half of the total membrane lipid exists in the gel
state, the temperature coefficient of growth increases rather
suddenly to values of 40-45 kcal/mole. If the proportion of
crystallized lipid rises still further to values in excess of
approximately 90%, cell growth ceases entirely. My interpretation of these observations is as follows: At temperatures
within the phase transition range, but above the midpoint of
the lipid phase transitions, a sufficiently high proportion
of the membrane lipids exists in a functional (i.e., a fluid)
state so that the normal growth rate characteristic of cells
whose membrane lipids are fully liquid-crystalline may be
maintained. The temperature coefficient of growth is solely
determined by temperature-dependence of one or more processes
occurring in those parts of the membrane that remain fluid.
As the environmental temperature falls below the midpoint of
the lipid phase transition, physiologically significant proportions of plasma membrane now become nonfunctional (solidified), thus not permitting the potential growth rate characteristic of predominantly liquid-crystalline cells to be
obtained. The temperature coefficient now increases because
both the total area of membrane remaining functional, as well
as the temperature-dependence of physiological processes
occurring with the remaining functional areas of the membrane,
become growth-limiting. Eventually, as the lower end of the
phase transition is approached, almost no functional, fluid
membrane regions remain, and cell growth ceases. This loss
of membrane function may well be due to conformational changes
in one or more integral membrane proteins, induced by a local
phase transition in the "boundary lipid" associated with
these proteins (Jost et al., 1973; Trauble and Overath, 1973),
or to defective membrane assembly at the lower boundary of
the phase transition, such as has been demonstrated recently
for Escherichia coli (Tsukagoshi and Fox, 1973). The rapid
decline of growth rate below the phase transition midpoint
may also be due, at least in part, to structural perturbations resulting from the accommodation of rigid, planar
arrays of crystallized lipid into small spherical cells.
Indeed, evidence for a transient breakdown in the permeability barrier apparently correlated with the lower end of the
membrane lipid phase transition has recently been reported
for A. laidlawii (McElhaney et al., 1973) and for a fatty
acid auxotroph of E. coli (Haest et al., 1972).

It is of interest to note that while the physical state

of the membrane has a definite effect on the absolute growth
rate at any given temperature, the maximum growth rates and
the growth yields obtained at the optimum growth temperature
did not vary significantly with the fatty acid composition of
the membrane lipids. Cells grown on stearic acid were the
only exception to these rules. With regard to the postulate
discussed above, that cells growing at temperatures above
their phase transition midpoint temperature have an "excess"
of functional, liquid-crystalline bilayer, it should be noted
that stearate-enriched cells are the only cells in this study
whose optimum growth temperature falls below their phase
transition midpoint. Thus stearic acid-grown cells, even at
their relatively high optimum growth temperature of 38°C,
still have the bulk of the membrane lipid existing in the gel
state. The suboptimal areas of functional, fluid membrane,
which may exist only in the case of cells supplemented with
stearic acid, could account for the reduced growth rate and
cell yield obtained at the optimum growth temperature of
these cells. Although at temperatures above 38°C the fraction of membrane lipid in the liquid-crystalline state
increases, the thermal inactivation of some cellular component rapidly becomes growth-limiting so that the potential
growth rate that is obtained with other cells containing an
"excess" of fluid membrane is never obtained in stearate-
enriched cells.

A number of recent studies support the idea that the
liquid-crystalline state is necessary to support growth
(Overath *et al.*, 1970), membrane transport (Overath *et al.*,
1970; Machtiger and Fox, 1973), and the activity of membrane-
associated enzymes (Kimelberg and Papahadjopoulos, 1972;
De Kruyff *et al.*, 1973). The results of a recent study of
the temperature-dependence of the activity of the membrane-
bound ATPase of A. laidlawii may be particularly pertinent to
the present study. De Kruyff *et al.* (1973) report an inflection in the Arrhenius plots of this enzyme, the temperature
at which inflection occurs depending upon the fatty acid composition of the membrane lipids, and correlating well with
the lower portion of the liquid-crystalline to gel membrane
lipid phase transition. The very rapid decline in ATPase
activity with temperature as the conversion to the gel state
nears completion, noted by these workers, may well be related
to the similar decline in the growth rate of this organism
described in the present study. It would be of interest to
study the effect of variations in the fatty acid composition
of the membrane lipids on the temperatures at which optimal
enzymic activity and thermal denaturation of this key

membrane-bound enzyme occur, to determine if these parameters might also correlate with the optimum and maximum growth temperatures of cells grown in the same fatty acids.

Souza *et al.* (1974) have isolated a temperature-sensitive mutant of a thermophilic bacillus that is unable to grow or to maintain the integrity of its membrane at high temperatures. It was found that this mutant also lacks the ability to alter its membrane lipid fatty acid composition in response to elevated temperature. From these and other data, these investigators concluded that the temperature-sensitivity of the mutant is probably a result of this defect in normal lipid metabolism at high temperatures, and that the changes in fatty acid composition normally observed at increased growth temperatures may be an essential feature of thermophily. The existence and behavior of such mutants provides additional evidence in favor of the hypothesis discussed earlier.

In summary, the results of many recent experiments support the concept that, at least under certain circumstances, the temperature range over which microbial growth can occur and the rate of growth within that temperature range can both be determined by the physical state of the membrane lipids. Contrary to the original suggestions of Heilbrunn (1924) and Bĕlehrádek (1931), the "melting point" of the cellular lipid determines the lower, not the upper, temperature limit of growth. It thus seems reasonable to conclude that the characteristic temperature-induced alterations in the fatty acid composition of microbial membranes does represent a mechanism which has evolved to ensure an optimal level of membrane lipid fluidity in a variety of different or fluctuating thermal environments. Such a homeostatic mechanism can account, at least in part, for the rather wide growth temperature ranges of many microorganisms and for the generally comparable rates of many membrane processes in microbes adapted for life at low, moderate, and high temperatures. However, changes in the physical state of the membrane lipids alone will not convert mesophiles such as A. laidlawii or Escherichia coli into psychrophilic or thermophilic organisms. Clearly, appropriate evolutionary changes in the structure and stability of certain membrane-associated and cytoplasmic macromolecules would also be required. Thus, the interpretation of these recent experiments as support for one "theory" of psychrophily or thermophily in exclusion to other theories would in my opinion, be unwarranted. Appropriate alterations in the physical state of the membrane lipids, induced by characteristic changes in fatty acid composition in response to temperature, probably represent a necessary, but certainly not sufficient condition for the existence of life in extreme thermal environments.

ACKNOWLEDGMENTS

This work was supported by a grant (MA-4261) from the Medical Research Council of Canada.

REFERENCES

BĚLEHRÁDEK, J. (1931). Protoplasma 12, 406.

BRETSCHER, M. S. (1973). Science 181, 622.

BROCK, T. D. (1967). Science 158, 1012.

CAPALDI, R. A. (1974). Scientific Amer. 230, 26.

CHAN, M., HIMES, R. H., and AKAGI, J. M. (1971).
    J. Bacteriol. 106, 876.

CHAPMAN, D. (1967). In "Thermobiology" (Ross, A. H., ed.),
    Academic Press, London.

DARON, H. H. (1970). J. Bacteriol. 101, 145.

DE GIER, J., MANDERSLOOT, J. C., HUPKES, J. V., McELHANEY,
    R. N., and VAN BEEK, W. P. (1971). Biochim.
    Biophys. Acta 233, 610.

DE KRUYFF, B., VAN DIJK, P. W. M., GOLDBACH, R. W., DEMEL,
    R. A., and VAN DEENEN, L. L. M. (1973). Biochim.
    Biophys. Acta 330, 269.

DEUEL, H. J. (1955). "The Lipids," Vol. 2, p. 586. Inter-
    science Publishers, New York.

DIAMOND, J. M., and WRIGHT, E. M. (1969). Ann. Rev.
    Physiol. 31, 581.

FOX, C. F., and KEITH, A. (1972). "Membrane Molecular
    Biology," pp. 164-228. Sinauer Associates, Inc.,
    Stamford, Conn.

FRAENKEL, G., and HOPF, H. S. (1940). Biochem. J. 34, 1085.

GAUGHRAN, E. R. L. (1947a). Bacteriol. Revs. 11, 189.

GAUGHRAN, E. R. L. (1947b). J. Bacteriol. 53, 506.

HAEST, C. W. M., DE GIER, J., VAN ES, G. A., VERKLEIJ, A. J., and VAN DEENEN, L. L. M. (1972). Biochim. Biophys. Acta 288, 43.

HEILBRUNN, L. V. (1924). Amer. J. Physiol. 69, 190.

HILDITCH, T. P. (1956). "The Chemical Constitution of Natural Fats," 3rd ed., pp. 171, 190. Chapman and Hall, London.

JOST, P. C., GRIFFITH, O. H., CAPALDI, R. A., and VANDERKOOI, G. (1973). Proc. Nat. Acad. Sci. U.S.A. 70, 480.

KATES, M., and BAXTER, R. M. (1962). Can. J. Biochem. Physiol. 40, 1213.

KATES, M., and HAGEN, P.-O. (1964). Can. J. Biochem. 42, 481.

KIMELBERG, H. K., and PAPAHADJOPOULOS, D. (1972). Biochim. Biophys. Acta 282, 277.

LEATHES, J. B., and RAPER, H. S. (1925). "The Fats," 2nd ed., pp. 100, 118. Longmans, Green and Co., London.

LUZZATI, V., and HUSSON, F. (1962). J. Cell. Biol. 12, 207.

MACHTIGER, N. A., and FOX, C. F. (1973). Ann. Rev. Biochem. 42, 575.

MARR, A. G., and INGRAHAM, J. L. (1962). J. Bacteriol. 84, 1260.

McELHANEY, R. N. (1974a). J. Mol. Biol. 84, 145.

McELHANEY, R. N. (1974b). J. Supramol. Struct. (in press).

McELHANEY, R. N. (1974c). PAABS Rev. (in press).

McELHANEY, R. N., and TOURTELLOTTE, M. E. (1969). Science 164, 433.

McELHANEY, R. N., DE GIER, J., and VAN DER NEUT-KOK, E. C. M. (1973). Biochim. Biophys. Acta 298, 500.

OLDFIELD, E., and CHAPMAN, D. (1972). FEBS Lett. 23, 285.

OVERATH, P., SCHAIRER, H. U., and STOFFEL, W. (1970). Proc. Nat. Acad. Sci. U.S.A. 67, 606.

PEARSON, L. K., and RAPER, H. S. (1927). Biochem. J. 21, 875.

PISETSKY, D., and TERRY, T. M. (1972). Biochim. Biophys. Acta 274, 95.

POLLACK, J. D., and TOURTELLOTTE, M. E. (1967). J. Bacteriol. 93, 636.

RAY, P. H., WHITE, D. C., and BROCK, T. D. (1971a). J. Bacteriol. 106, 25.

RAY, P. H., WHITE, D. C., and BROCK, T. D. (1971b). J. Bacteriol. 108, 227.

REISS-HUSSON, F., and LUZZATI, V. (1967). Adv. Biol. Med. Phys. 11, 87.

SHEN, P. Y., COLES, E., FOOTE, J. L., and STENESH, J. (1970). J. Bacteriol. 103, 479.

SINGER, S. J., and NICOLSON, G. L. (1972). Science 175, 720.

SINGLETON, R., and AMELUNXEN, R. E. (1973). Bacteriol. Rev. 37, 320.

SMITH, P. F. (1971). "The Biology of Mycoplasmas," pp. 6-41. Academic Press, New York.

SOUZA, K. A., KOSTIW, L. L., and TYSON, B. J. (1974). Arch. Microbiol. 97, 89.

STEIM, J. M., TOURTELLOTTE, M. E., REINERT, J. C., McELHANEY, R. N., and RADER, R. L. (1969). Proc. Nat. Acad. Sci. U.S.A. 63, 104.

STEIN, W. D. (1967). "The Movement of Molecules across Cell Membranes," pp. 65-105. Academic Press, New York.

TERROINE, E. F., BONNET, R., KOPP, G., and VECHOT, J. (1927). Bull. Soc. Chim. Biol. 12, 605.

TERROINE, E. F., HATTERER, C., and ROEHRIG, P. (1930). Bull. Soc. Chim. Biol. 12, 682.

TRAUBLE, H., and OVERATH, P. (1973). Biochim. Biophys. Acta
307, 491.

TSUKAGOSHI, N., and FOX, C. F. (1973). Biochemistry 12, 2816.

WISDOM, C., and WELKER, N. E. (1973). J. Bacteriol. 114, 1336.

GROWTH TEMPERATURE AND THE STRUCTURE OF THE LIPID PHASE IN BIOLOGICAL MEMBRANES

Alfred F. Esser and Kenneth A. Souza

During the past few years, intensive research efforts have been made toward the understanding of structure-function relationships in biological membranes. The general consensus seems to be that the structure of functional membranes is a kind of two-dimensional arrangement of proteins embedded in a lipid bilayer (Stoeckenius and Engelman, 1969; Singer, 1971; Wallach, 1972). Although the bilayer concept appears to be valid for all membranes, a great diversity exists in the detailed disposition of proteins in the bilayer. There are certain regions within cell membranes in which proteins are (a) in a crystalline array, as in the purple membrane of Halobacterium halobium (Blaurock and Stoeckenius, 1971), or the cytochrome oxidase-containing membrane (Jost et al., 1973), and the gap junctions of hepatocytes (Goodenough, 1972), or (b) are clustered, as was found, e.g., in Escherichia coli cell membranes (Kleeman and McConnell, 1974), or (c) apparently dispersed randomly as in erythrocyte membranes (Tillack and Marchesi, 1970). This variation may be due to specifically oriented protein-protein interaction or may be the consequence of the physical state of the lipids and the lipid-protein ratio in the membrane. It is certainly a reflection of the many different functions that are associated with cell membranes.

A unique property of the lipids in the bilayer is their ability to undergo a reversible, thermotropic gel to liquid-crystalline phase transition that arises from a cooperative melting of the hydrocarbon chains in the interior of the bilayer. Thus, the physical state of lipids in membranes is strongly influenced by the environmental temperature, and a vast number of techniques have been used to study the phase transition and the dynamics of lipids in membranes (Oldfield and Chapman, 1972). The temperatures at which phase changes occur are dependent upon the head-group, the length, the degree and type of unsaturation of the hydrocarbon chain, and the amount of branching. However, lipid phase transitions can be abolished by certain substances such as cholesterol (Oldfield and Chapman, 1972), or lysolecithin (Keith *et al.*, 1975), and through the influence of proteins (Esser and Lanyi, 1973). The reason for the suppression is at the moment poorly understood (Oldfield and Chapman, 1972), but may have to do with the fact that for the transition to occur, a "cooperative unit" in the order of 70 molecules must be possible (Scott, 1974).

A consequence of this "rigid-to-fluid" transition that is of prime importance for membrane function is the fact that in the fluid state the lipids as well as the proteins can move rapidly in the plane of the membrane (Scandella *et al.*, 1972; Poo and Cone, 1974). McConnell and his associates have also pointed out the difference between phase changes in single component systems and phase separations in multicomponent systems, and the relevance of the latter to observed functions in biological membranes (Shimshick and McConnell, 1973; Linden *et al.*, 1973; Kleeman and McConnell, 1974). Furthermore, it is well established that phase changes are related to changes in the activity of some membrane-bound enzymes (Lyons, 1972), growth rates (McElhaney, 1974), and cell lysis (Steim *et al.*, 1969).

The most convincing evidence for such relationships was obtained from studies with Acholeplasma laidlawii membranes. The membrane lipids of this organism can be varied considerably by growing the cells on defined media supplemented with various fatty acids (McElhaney, this volume). The fatty acid composition of certain β-oxidation-deficient fatty acid auxotrophs of E. coli can be varied in a similar manner (Overath *et al.*, 1971), and for both organisms it was found that the cells would not grow when the lipids were in an "all-frozen" state (Machtiger and Fox, 1973).

A more natural change in the fatty acid composition of
membrane lipids can be brought about by a change in growth
temperature. Microorganisms grown at increasing temperatures
tend to incorporate more saturated, less branched, and longer
fatty acid chains into their phospholipids (Cronin and
Vagelos, 1972). As a result, one would expect that phase
transitions and phase separations of phospholipids in these
membranes should occur at progressively higher temperatures.
These processes thus may be of great importance for the survival of organisms and their adaptation to higher temperatures.

The molecular mechanisms that allow thermophilic bacteria to grow at high temperatures are poorly understood at
present (Singleton and Amelunxen, 1973). Even less is known
about the factors that limit growth at certain temperatures.
We attempt in this paper to develop a perspective on some of
the requirements that have to be met in order for high temperature adaptation to occur. We will limit the scope of our
discussion to lipids and membranes, since other cell components of importance, such as proteins and nucleic acids, are
discussed elsewhere in this volume.

Our arguments are based mainly on our own studies on
Bacillus stearothermophilus (Souza et al., 1974; Esser and
Souza, 1974), those of Sinensky (1974), Overath et al. (1971),
and Tsukagoshi and Fox (1973) on E. coli, and those of
McElhaney (1974) and Huang et al. (1974) on A. laidlawii. A
direct comparison of the published results is facilitated by
the fact that they were obtained mainly with two methods,
thermal analysis (differential thermal analysis, DTA; and
differential scanning calorimetry, DSC) and the electron
paramagnetic resonance (EPR) spin-labeling technique. A
detailed description of both methods including their advantages and pitfalls can be found in excellent reviews
(Ladbrooke and Chapman, 1969; Smith, 1972; Keith et al.,
1973; Wallach and Winzler, 1974) and, therefore, will not be
included here.

HOMEOVISCOUS ADAPTATION

A most fascinating and interesting result that has
emerged from spin-label studies on the three microorganisms
mentioned above is the fact that the labels always experience
the same amount of motional restriction at different growth
temperatures. In other words, B. stearothermophilus,
A. laidlawii, and E. coli are able to produce membranes whose

lipids have a constant fluidity at the temperature of growth
--a process that Sinensky (1974) calls "homeoviscous adaptation". It is interesting to note that the viscosities of the
three different membranes are apparently very similar. When
the anisotropy parameter $2A_m$ (or $2T_{//}$) is used to represent
membrane fluidity a value of 51-52 Gauss is found for
B. stearothermophilus (Esser and Souza, 1974) and of 53-54
Gauss for A. laidlawii (Huang et al., 1974), or when the
rotational correlation time of the appropriate spin-label is
used to calculate viscosities then values of 2.5 poise for
E. coli (Sinensky, 1974) and 2.2 poise for B. stearothermophilus (Esser, unpublished) are obtained.

The principle of homeoviscous adaptation seems to be
established fairly well for these cells, but it is less clear
which membrane-specific function would require such a strict
conservation of viscosity. It is well documented that certain enzyme activities are lipid phase dependent; examples
are ATPase in A. laidlawii (de Kruyff et al., 1973) and
β-galactosidase (Tsukagoshi and Fox, 1973) in E. coli. But
other membrane-bound enzymes in the same organisms show no or
very little lipid phase dependence; examples are NADH oxidase
and p-nitrophenyl phosphatase in A. laidlawii (de Kruyff
et al., 1973) and phosphotransferase activity (Wilson et al.,
1970) and glycerol-3-phosphate dehydrogenase (Mavis and
Vagelos, 1972) in E. coli. It could be that homeoviscosity
is necessary for membrane biogenesis or cell division, but
clearly more experiments aimed at this question are required
before we can ascertain the necessity for homeoviscous
adaptation.

MINIMAL GROWTH TEMPERATURE

The second important general result that emerged from
the spin-label and calorimetry studies on the three microorganisms is the fact that a certain minimum amount of fluid
lipids is necessary for growth. McElhaney (1974) has shown
that if the proportion of solid lipids rises above 90%, cell
growth of A. laidlawii ceases entirely. Similar results
were reported by Overath et al. (1971) and Tsukagoshi and
Fox (1973) for E. coli. Organisms will make every effort to
avoid such conditions and for microorganisms the most effi-
cient way to do this is to change the fatty acid lipid composition. A very ingenious way to cope with this problem apparently functions in some hibernating mammals. Hibernating
or heterothermic organisms normally maintain body temperatures of above 37°C, but can temporarily abandon the

homeothermic state, lower their temperature to about 10°C, and still maintain all vital membrane functions (Hensel *et al.*, 1973). According to recent data by Keith *et al.* (1975) one mechanism used by the hibernating ground squirrel is the incorporation of lysophosphatides into membranes, thereby preventing solidification of the lipids.

In summary, we can state quite safely that in organisms whose membranes show phase changes, the presence of all lipid hydrocarbon chains in the solid state is incompatible with growth. This does not imply, of course, that the minimum growth temperature could not fall above the phase transition temperature; temperature dependent changes in proteins and nucleic acids could likewise determine the minimal growth temperature.

MAXIMAL GROWTH TEMPERATURE

Through the extensive work of Brock, it is known that there are certain temperature limits up to which certain kinds of organisms can survive and multiply. Thus, he found that eukaryotic organisms will not grow above 60°C (Tansey and Brock, 1972), while prokaryotic organisms can grow up to 95°C (Brock, 1967), a notable exception being prokaryotic, photosynthetic blue-green algae which are not found to survive above 75°C (Brock, 1967). The reasons for such temperature limits are unknown and only speculative theories have been advanced thus far. Tansey and Brock (1972), for example, suggested that the thermostability of cell membranes might be a factor in limiting growth at higher temperatures. In the case of eukaryotic cells they state specifically that membranes of cell organelles (mitochondria, nuclei, etc.) might be the limiting structures; however, no molecular explanation or experimental results were provided. McElhaney (1974) in his excellent report on membrane lipids and growth temperature in A. laidlawii found no definite and constant relationship between the maximal growth temperature and the physical state of membrane lipids, although he argues that "apparently an upper, as well as a lower, limit exists on the membrane lipid fluidity that is compatible with growth... ." This is reminiscent of the older theories of Heilbrunn (1924) and Bĕlehrádek (1931), which state that the maximal growth temperature is limited by the melting points of lipids. Yet, Babel *et al.* (1972) argue that an exact relationship between maximal growth temperature and melting points cannot be demonstrated.

We have proposed (Esser and Souza, 1974) that the temperature limit for thermophilic growth may be determined by the boundary conditions of the phase diagram for the total lipid mixture in the membrane. In other words, as long as there are two lipid phases present in the membrane at the growth temperature the cells can survive and divide. This hypothesis is based largely on results obtained with mutant TS-13 of B. stearothermophilus. This mutant fails to grow above 58°C, whereas the wild type can grow up to 72°C (Souza et al., 1974). Our spin-label data indicate that the lipids in mutant membranes grown at either 52°C or 58°C undergo phase separations at the same temperatures. This is not surprising since the fatty acid composition of the membrane lipids in the differently grown cells is very similar (Table 1). In both cell types the phase separation starts at 58°C; above that temperature all lipids are in the fluid state. Since 58°C is also the maximal growth temperature for the mutant, we interpret these results to indicate that B. stearothermophilus cells cannot grow above the temperature for phase separations and that this temperature determines the maximal growth temperature.

Our hypothesis is supported partially by spin-label data obtained with A. laidlawii by Huang et al. (1974), but is not easily reconciled with calorimetric data of McElhaney (1974) on the same organism. Huang et al. (1974) found that cells grown at 28°C and on different fatty acids possess upper phase changes at 40°-43°C which is very close to the maximal growth temperature of 44°C for this organism (McElhaney, 1974). On the other hand, McElhaney's DTA data indicate that A. laidlawii cells can grow far above the phase transition range observed by thermal analysis. We have observed likewise, that in B. stearothermophilus lipids the phase transitions measured by DSC are completed below the growth temperature (Souza, unpublished). We feel that these differences in transition temperatures are due to the different methods employed to study them. Why do the spin-label and thermal analysis techniques report different transition temperatures? We would like to offer the following answer to this question.

DTA or DSC record bulk changes in the hydrocarbon phase of lipid mixtures, and there is no doubt that all hydrocarbon chains are in the gel state below the transition and all in the liquid-crystalline state above the transition. Within the transition range both phases exist simultaneously (Oldfield and Chapman, 1972). But only in rare cases is it

288

Table 1

Fatty Acid Composition of Mutant TS-13[a]

Fatty acid methyl esters[b]	52°C	58°C
u.i.[c]	T[d]	T[d]
14:0i	2.4	2.5
14:0	4.2	4.5
15:0i	13.2	17.4
15:0a	3.9	4.0
15:0	4.3	3.8
16:0i	22.6	19.7
16:0a	T	T
16:0	20.0	20.8
16:1	6.6	2.6
17:0i	6.9	8.6
17:0a	10.8	12.8
17:0	1.9	1.8
u.i.	T	T
18:0	2.1	2.0
18:1	T	T
u.i.	T	T

[a]Values are expressed as per cent of total fatty acids. Averages of at least four determinations representing two different cell batches and two analyses of each are reported.

[b]Number of carbon atoms:number of double bonds; i-iso; a-anteiso.

[c]u.i. = unidentified.

[d]T = trace, < 1.0%.

possible to record phase separations with thermal analysis techniques. The spin-labels, however, have the potential to be located predominantly in one or the other phase, and thus can report phase separations (Shimshick and McConnell, 1973) directly. Depending on what type of label is used one can, in theory, observe separations of the hydrocarbon region and/or the head-group region, but one could also observe mixing and demixing of different lipid classes within one phase. Partial miscibility can be temperature dependent, and the temperature at which mixing occurs may or may not be equivalent to one of the transition temperatures. Thus, in a multicomponent system it is possible to have mixed or separated components present in the same phase depending on the temperature (see phase diagrams in textbooks of physical chemistry, e.g., Sheehan, 1970). It is conceivable that the spin-label method, especially when a "head-group label" is used, will report changes in miscibilities of a multicomponent system rather than solid to fluid transitions or separations. Obviously, changes of this type are difficult to observe by thermal analysis, and this might explain the different "transitions" that are recorded with the two methods. It becomes necessary to distinguish clearly between phase separations that depend upon composition from those that depend upon the state of the matter, i.e., mixing and demixing versus fluid-solid transitions.

The possibility that the changes that we have observed with spin-labels in B. stearothermophilus membranes may be due to mixing and demixing of lipids does not invalidate our hypothesis, but rather broadens its scope. We have postulated that as long as there are two lipid phases present in the membrane at the growth temperature, the cells can survive. The presence of two phases, e.g., gel and liquid-crystalline, ensures the formation of domains containing predominantly one class of lipids. A similar situation would occur if lipids are partially immiscible at the growth temperature, despite the fact that they may be all in the fluid state. Of course, the question that has to be asked now is: why is it necessary that certain domains enriched in one kind of lipid should exist? And why would complete randomization be incompatible with growth and reproduction?

Several answers to these questions have been proposed already. For example, Rothfield and Romeo (1971) have suggested that during membrane biogenesis a newly synthesized, lipid-requiring enzyme has to be inserted directly into portions of membranes having a specific lipid domain.

Furthermore, considering that some membrane-associated functions require an ordered arrangement of enzymes, for example the electron transport chain in the inner mitochondrial membrane (Capaldi and Green, 1972; Harmon et al., 1974), it is very likely that randomization would interfere with function.

In summary, the assertion has been made that the ability of microorganisms to adjust their membrane lipid composition in such a way as to produce constant micro-phase-separations in response to changes in growth temperature constitutes a prime mechanism for high (or low) temperature adaptation. Due to lack of specific data our hypothesis does not consider any effects that changes in protein composition may have on membrane phase separations. Yet, it is likely that alterations of the charge or conformation of proteins will affect miscibility, and the results obtained by Hubbell and Blasie and their coworkers on reconstituted and native rhodopsin membranes clearly point in this direction (Hong and Hubbell, 1973; Chen and Hubbell, 1973; Blasie, 1972). In the future it will be necessary to initiate studies on growth temperature-dependent changes in proteins (and nucleic acids?) not only to learn more about their inherent thermostability, as has been done so far, but also to investigate possible consequences on membrane architecture.

ACKNOWLEDGMENTS

The authors are indebted to Drs. D. Chapman and R. McElhaney for criticisms and comments, and to Dr. R. Aloia for providing data prior to publication. A. F. E. thanks the Research Corporation and NASA's Ames Research Center (NCAR-253-501) for financial support. This is publication No. 19 from the Institute for Molecular Biology, California State University, Fullerton.

REFERENCES

BABEL, W., ROSENTHAL, H. A., and RAPOPORT, S. (1972). Acta
    Biol. Med. Ger. 28, 565.

BĚLEHRÁDEK, J. (1931). Protoplasma 12, 406.

BLASIE, J. K. (1972). Biophys. J. 12, 205.

BLAUROCK, A. E., and STOECKENIUS, W. (1971). Nature New Biol.
    233, 152.

BROCK, T. D. (1967). Science 158, 1012.

CAPALDI, R. A., and GREEN, D. E. (1972). FEBS Lett. 25, 205.

CHEN, Y. S., and HUBBELL, W. L. (1973). Exp. Eye Res. 17, 517.

CRONIN, J. E., and VAGELOS, P. R. (1972). Biochim. Biophys. Acta 265, 25.

DE KRUYFF, B., VAN DIJK, P. W. M., GOLDBACH, R. W., DEMEL, R. A., and VAN DEENEN, L. L. M. (1973). Biochim. Biophys. Acta 330, 269.

ESSER, A. F., and LANYI, J. K. (1973). Biochemistry 12, 1933.

ESSER, A. F., and SOUZA, K. A. (1974). Proc. Nat. Acad. Sci. U.S.A. 71, 4111.

GOODENOUGH, D. A. (1972). In "Membrane Research" (C. F. Fox, ed.), p. 337, Academic Press, New York.

HARMON, H. J., HALL, D. J., and CRANE, F. L. (1974). Biochim. Biophys. Acta 344, 119.

HEILBRUNN, L. V. (1924). Amer. J. Physiol. 69, 190.

HENSEL, H., BRUCK, K., and RATHS, P. (1973). In "Temperature and Life" (H. Precht, J. Christopherson, H. Hensel, and W. Larcher, eds.), p. 505, Springer Verlag, Berlin.

HONG, K., and HUBBELL, W. L. (1973). Biochemistry 12, 4517.

HUANG, L., LORCH, S. K., SMITH, G. S., and HAUG, A. (1974). FEBS Lett. 43, 1.

JOST, P. C., GRIFFITH, O. H., CAPALDI, R. A., and VANDERKOOI, G. (1973). Proc. Nat. Acad. Sci. U.S.A. 70, 480.

KEITH, A. D., SHARNOFF, M., and COHN, G. E. (1973). Biochim. Biophys. Acta 300, 379.

KEITH, A. D., ALOIA, R. C., LYONS, J., SNIPES, W., and PENGELLEY, E.T. (1975). Biochim. Biophys. Acta 394, 204.

KLEEMAN, W., and McCONNELL, H. M. (1974). Biochim. Biophys. Acta 345, 220.

LADBROOKE, B. D., and CHAPMAN, D. (1969). Chem. Phys. Lipids 3, 304.

LINDEN, C. D., WRIGHT, K. L., McCONNELL, H. M., and FOX, C. F. (1973). Proc. Nat. Acad. Sci. U.S.A. 70, 2271.

LYONS, J. M. (1972). Cryobiology 9, 341.

MACHTIGER, N. A., and FOX, C. F. (1973). Ann. Rev. Biochem. 42, 575.

MAVIS, R. D., and VAGELOS, P. R. (1972). J. Biol. Chem. 247, 652.

McELHANEY, R. N. (1974). J. Mol. Biol. 84, 145.

OLDFIELD, E., and CHAPMAN, D. (1972). FEBS Lett. 23, 285.

OVERATH, P., SCHAIRER, H.-U., HILL, F. F., and LAMNEK-HIRSCH, I. (1971). In "The Dynamic Structure of Cell Membranes" (D. F. H. Wallach, and H. Fischer, eds.), p. 149, Springer Verlag, Berlin.

POO, M.-M., and CONE, R. A. (1974). Nature 247, 438.

ROTHFIELD, L., and ROMEO, D. (1971). In "Structure and Function of Biological Membranes" (L. I. Rothfield, ed.), p. 251, Academic Press, New York.

SCANDELLA, C. J., DEVAUX, P., and McCONNELL, H. M. (1972). Proc. Nat. Acad. Sci. U.S.A. 69, 2056.

SCOTT, H. L., Jr. (1974). J. Theor. Biol. 46, 241.

SHEEHAN, W. F. (1970). "Physical Chemistry" 2. Ed. pp. 312-322, Allyn and Bacon, Boston.

SHIMSHICK, E. J., and McCONNELL, H. M. (1973). Biochemistry 12, 2351.

SINENSKY, M. (1974). Proc. Nat. Acad. Sci. U.S.A. 71, 522.

SINGER, S. J. (1971). In "Structure and Function of Biological Membranes" (L. I. Rothfield, ed.), p. 145, Academic Press, New York.

SINGLETON, R., Jr., and AMELUNXEN, R.E. (1973). Bacteriol.
Rev. 37, 320.

SMITH, I. C. P. (1972). In "Biological Applications of Electron Spin Resonance" (H. M. Swartz, J. R. Bolton,
and D. C. Borg, eds.), Wiley-Interscience, New York.

SOUZA, K. A., KOSTIW, L. L., and TYSON, B. J. (1974). Arch.
Microbiol. 97, 89.

STEIM, J. M., TOURTELLOTTE, M. E., REINERT, J. C., McELHANEY,
R. N., and RADER, R. L. (1969). Proc. Nat. Acad.
Sci. U.S.A. 63, 104.

STOECKENIUS, W., and ENGELMAN, D. M. (1969). J. Cell Biol.
42, 613.

TANSEY, M. R., and BROCK, T. D. (1972). Proc. Nat. Acad. Sci.
U.S.A. 69, 2426.

TILLACK, T. W., and MARCHESI, V. T. (1970). J. Cell. Biol. 45,
649.

TSUKAGOSHI, N., and FOX, C. F. (1973). Biochemistry 12, 1973.

WALLACH, D. F. H. (1972). "The Plasma Membrane," Springer
Verlag, New York.

WALLACH, D. F. H., and WINZLER, R. J. (1974). "Evolving
Strategies and Tactics in Membrane Research,"
Springer Verlag, New York.

WILSON, G., ROSE, S. P., and FOX, C. F. (1970). Biochem.
Biophys. Res. Commun. 38, 617.

Note added in proof: While this paper was in the editorial
process a review by G. Vanderkooi appeared (Biochim. Biophys.
Acta 344, 307) that expressed similar ideas on fluid phase
separations of lipids and their importance for membrane
function. Furthermore, Wu and McConnell (Biochemistry 14, 847)
reported direct experimental evidence for fluid-fluid
immiscibility of lipids in model membranes.

MEMBRANE STRUCTURE AND SALT DEPENDENCE IN EXTREMELY HALOPHILIC BACTERIA

Janos K. Lanyi

Halobacterium cutirubrum and other halobacteria show
optimal growth in the presence of 3.5-5.0 M NaCl. When the
salt concentration is lowered, dramatic changes are seen:
the cell first breaks open (Abram and Gibbons, 1961) and the
cell envelope becomes fragmented. The fragments disintegrate
to give a clear solution at NaCl concentrations below
0.1-0.2 M (Onishi and Kushner, 1966; Brown, 1963; Stoeckenius
and Rowen, 1967; Lanyi, 1971). Isolated cell envelope vesicles
behave similarly: the vesicles first become permeable (at
2 M NaCl), an outer layer of protein particles is lost at
1 M NaCl, and upon removal of the residual salt only small
membrane fragments remain (Stoeckenius and Rowen, 1967;
Lanyi, 1971).

There are numerous technical problems in investigating
these salt-dependent phenomena. Ideally, a description of
the effect of salt on these membranes should be couched in
terms of residue interactions between polypeptide chains,
lipid matrix stability, and tendencies for lipid-protein
binding. The cell envelope of extreme halophiles, similarly
to other membranes, is a complex entity, however, consisting
of many different proteins (Brown and Pearce, 1969; Mescher
et al., 1974) and lipids (Tornabene et al., 1969; Kates, 1972;
Kushwaha et al., 1972; Kushwaha et al., 1974). Only limited

studies have been carried out involving but a few of these components, thus, they are necessarily incomplete and can only indicate in broad outline the possible solutions to the question of salt response in halophilic membranes.

The enzyme activities of membrane proteins in halophilic systems have been studied for many years (Larsen, 1967; Lanyi, 1974a). Membrane enzymes of halophiles, as exemplified by components of the electron transport chain, nearly always show little activity in the absence of salt, and attain maximal activity at several molar salt concentrations. In addition, these enzymes exhibit a time-dependent loss of activity not restored by addition of salt, when incubated at low ionic strengths (Lanyi, 1969a, b; Lieberman and Lanyi, 1971). Salt influences charge inter-actions in proteins in general, and since many halophilic proteins have been found to contain an excess of acidic amino acids (Stoeckenius and Kunau, 1968; Reistad, 1970; Visentin *et al.*, 1972), the proposal that the salt effects involve the screening of charges (Baxter, 1959) is reasonable.

The first clues that other types of interactions are also involved originated from the observations that the salt stabilization of halophilic enzymes was highly anion specific and that the salts employed fit the lyotropic series, with salting-out type salts being the most effective (Lanyi and Stevenson, 1970). A variety of other approaches with several halophilic enzymes, including study of the thermodynamics of denaturation and the use of various selective protein denaturants, indicated that the predominant function of salt was to stabilize weak hydrophobic inter-actions which are apparently not sufficiently stable in water alone (Lanyi, 1974a).

It seems from the above that at least some proteins in the membranes of the halophiles must undergo conformational changes after removal of salt to the extent that they lose their functional properties. Direct measurements of titratable groups (Brown, 1965) and PMR line-broadening (Chapman and Kamat, 1968) of H. cutirubrum membrane proteins also indicated that unfolding of the polypeptide chains takes place at lowered NaCl concentrations. It is not clear, however, how such changes would affect the overall structure of the membranes. According to a generally accepted model for the architecture of biological membranes (Singer and Nicolson, 1972), structural stability originates from the lipid bilayer. In this model, proteins are placed either on the outer surfaces of the bilayer, attached by ionic forces, or inserted to various degrees into the bilayer in intimate contact with the hydrophobic core. Purely ionic interactions are susceptible to disruption by salts; hence it is unlikely that they have a predominant role in halophilic membranes. Hydrophobic lipid-protein bonds, on the other hand, are good candidates for mediating the stabilization of these membranes by high concentrations of salt. Indeed, anion selectivity, such as observed for halophilic enzymes, was also found in studies of the salt dependence of the integrity of halophilic membranes (Lanyi, 1971). In these studies the light-scattering of cell envelope preparations was seen to decrease in a discontinuous fashion as NaCl concentration was lowered. Furthermore, analyses of sedimentable protein and lipid phosphorus indicated that at NaCl concentrations above 1 M the disruption of the membranes resulted in the solubilization of proteins, whereas below this concentration the detachment of lipoproteins/membrane fragments began to occur. Anion specificity was observed for the solubilization of proteins only. This type of evidence and others (Lanyi, 1971), suggested that those components of the cell envelopes which become solubilized between 1-3 M NaCl (included in this category are some electron transport components) are attached primarily by weak hydrophobic bonds. It is possible that the loss of these proteins (comprising about 30% of the total envelope proteins) could cause some disorganization of the membranes. At lower concentrations of NaCl (< 1 M) the nega-tive charges on both proteins and lipids (Kates and Hancock, 1971) are expected to be shielded insufficiently; thus, the disintegration of the membranes into small (but still mem-brane-like) lipoprotein entities is probably due to charge repulsion. Protection against this latter type of disinte-gration by various salts is not anion-specific (Lanyi, 1971) and can be provided also by lower concentrations of divalent cations.

The lipids of halobacteria, which in these membranes, as in others, must contribute some of the structural stability, are unusual in the following respects:
    (1) All the polar lipids contain ether rather than ester linkages,
    (2) All the polar lipids contain a single kind of hydrocarbon chain (dihydrophytyl) with four methyl branches (Kates *et al.*, 1965),
    (3) The major polar lipid, phosphatidyl glycerophos-phate, has two ionized phosphate groups at neutral pH (Kates and Hancock, 1971; Kates, 1972),

(4)   There is a substantial fraction of non-polar lipids (8-10% by weight), consisting mainly of bacterioruberin (Kelly *et al.*, 1970) and squalenes (Tornabene *et al.*, 1969; Kramer *et al.*, 1972) in about equal amounts.

The possibility exists that some of the peculiarities in the chemistry of these lipids contribute to the observed salt-dependent effects in halophilic membranes either by modifying lipid-lipid interactions in such a way as to make them dependent on high salt concentrations for stability, or by restricting the lipid-protein interactions to those which are stable only at high salt concentrations.   The first possibility is unlikely, since vesicles prepared by dispersing H. cutirubrum lipids in aqueous solutions were found to remain osmotically intact only at low salt concentrations (Chen *et al.*, 1974), suggesting that the properties of the whole membranes reflect the influence of the membrane proteins.

Artificial membranes, prepared from lipids only, have been used as models of biological membranes.   The motion of unbranched lipid chains in such bilayers consists of the bending of C--C bonds, which occurs at increasing frequency toward the middle of the bilayer.   While this is undoubtedly true of H. cutirubrum polar lipids as well, for stereochemical reasons the loci of methyl-branching are expected to be preferred places for C--C bending (Plachy *et al.*, 1974).   This prediction is consistent with a number of observations (Plachy *et al.*, 1974), including segmental volume versus temperature plots and the temperature-dependent movement of spin-labeled stearic acid in these bilayers.   If the dynamics of chain motions is indeed dominated by bending at the four branching points, the bilayer should be a highly cooperative structure, with no distinct liquid-to-liquid crystalline phase transition as is prevalent in straight-chain systems.   Instead, motional freedom should be acquired by successive four-carbon segments of the lipid chain in a sequential fashion beginning in the middle of the bilayer, as temperature is increased. Indeed, no first-order phase transition was observed above -15°C, and the expansion coefficient of the lipid dispersions was found to change rapidly at four temperature points (Plachy *et al.*, 1974).

This kind of highly structured bilayer, suggested for the lipids of halophiles, would have some predicted consequences for the membranes of these organisms.   First, the insertion of hydrophobic proteins would produce more immobilization of these branched lipid chains than in straight-chain systems because the increased cooperativity between chains would extend the "boundary lipid" layer (Jost *et al.*, 1973) around the proteins to several chain diameters. In fact, in an earlier study with spin-labeled fatty acid probes (Esser and Lanyi, 1973), a comparison of H. cutirubrum membranes with dispersions of lipids from the same organism revealed extensive immobilization of the hydrophobic phase in the membranes by the proteins. An important consequence of this effect would be a preference for channel-type transport mechanisms over mobile carriers in halophilic membranes. We found that in H. halobium cell envelope vesicles, gramicidin (a channel-type carrier) is several orders of magnitude more effective as a cation carrier than is valinomycin (a mobile carrier), as expected (MacDonald and Lanyi, in preparation).

A second consequence of the proposed lipid structure in halophilic membranes is a decreased possibility for the steric accommodation of protein surfaces in the hydrophobic region of the bilayers. Thus, a lessened tendency for overlapping between the lipid chains and the hydrophobic residues of the membrane proteins, due to the restriction on the lipid chain motions imposed by the methyl branches, may contribute to the requirement for high concentrations of salt for hydrophobic stabilization. Information on specific membrane components is scarce, however, and the validity of this idea is difficult to assess.

We have suspected that squalene and its derivatives, which are present in halophilic membranes in relatively large amounts (Tornabene *et al.*, 1969), have some influence over the structure of the lipid phase. These compounds are long chain (C_{24} backbone) hydrocarbons, containing nonconjugated double bonds, which can nearly span the width of the hydrophobic core of the bilayer. Studies of the partitioning of a small amphiphilic spin label, di-tert-butyl nitroxide, between water and the lipid phase, revealed that the inclusion of squalene in the lipid dispersions displaced the distribution of the probe in favor of the aqueous phase (Lanyi *et al.*, 1974). According to our model of these membranes, discussed above, the solubility of the probe in the lipid phase is dependent on the size and number of the spaces and gaps that arise from the cooperative bending motions of the lipid chains. The results of the partitioning experiments suggest then, that squalene projects into both halves of the bilayer, perpendicularly to the plane of the membrane, and occupies some of the free lattice space.

Results obtained with a small fluorescent probe,

299

perylene, are in accord with the hypothesis above. Owing to
its hydrophobic character, this probe is expected to be found
in the lipid phase, even under very unfavorable conditions.
Studies of the free energy of activation for the thermal
motion of the probe indicated, however, that at a 6:1 molar
ratio of polar lipids to squalene the motion of perylene is
more dependent on the motions of the hydrocarbon chain lattice
than above or below this ratio (Lanyi *et al.*, 1974). The
suggestion was made, therefore, that at this molar ratio an
association between the lipids causes the formation of a
quasi-complex which reduces the free space in the bilayer.

Another observed consequence of the presence of squalene
in these lipid bilayers was its influence on the interaction
of the lipids with divalent cations. Previously it was noted
that the inclusion of the non-polar lipids of H. cutirubrum
caused a Mg^{2+}-dependent thermal event which resulted in some
kind of breakdown in the bilayer (Lanyi, 1974b). More
detailed studies indicated that the effect is analogous to
the divalent cation-induced aggregation of phospholipids in
general (Abramson *et al.*, 1964; 1968), but in the case of the
halophilic system the aggregation requires the presence of
squalene, at or above a 1:6 molar ratio to the polar lipids
(Lanyi *et al.*, 1974). The aggregation occurred suddenly, at
a temperature which was dependent on the squalene-polar lipid
ratio, on the concentration and nature of the divalent cation
(Ca^{2+} or Mg^{2+}), on the presence of NaCl or KCl (in the molar
concentration range), and on the pH. The results were consistent with the idea that the close packing of the polar
lipids (mainly consisting of phosphatidyl glycerophosphate)
prevents the access of cations to the inner layer of charged
phosphate groups, and that the effect of squalene is to
increase headgroup spacings, thereby making this region more
permeable (Figure 1). NaCl or KCl, as mentioned above, con-
tributed to the effect of divalent cations. In fact, the
influence of squalene on the behavior of the lipids was
observed mostly when conditions closely resembled the physiological milieu of halophiles: several molar NaCl, 100 mM
$MgCl_2$, temperatures of 30-40°, and squalene content of 5-10%
(relative to the total lipid). It is possible, therefore,
that local interactions between membrane proteins and the
lipid headgroups are also influenced, in a similar manner, by
the presence of squalene, and through this effect by high
salt concentrations as well.

From the above, it can be seen that the behavior of the
membranes of halophiles is very complex and has nurtured many

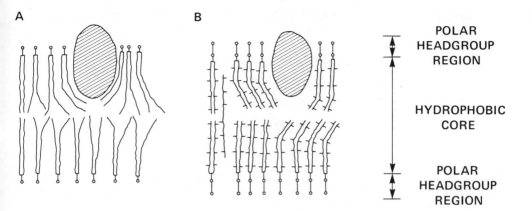

Fig. 1. *Schematic representation of bilayer containing straight-chain lipids (A), and phytanyl chain phosphatidyl glycerophosphate lipids (B). The cross-hatched shapes represent hypothetical membrane proteins which project partially into the hydrophobic phase. As indicated in (A), the unbranched chains are flexible and follow the contours of the protein. In membranes of halophilic bacteria, however, it is proposed that the chains are more rigid and show preferential bending at the branching points (B). In the latter illustration the mode of insertion for squalene (straight chain on left) is illustrated. From the sketch it can be seen how squalene might make the headgroup region more accessible to cations.*

hypotheses. It is likely that many factors contribute to the observed salt-dependence--a consequence of the structural complexity and the heterogeneity of the membranes. Further progress in understanding these effects is necessarily predicated on isolating single, well-defined membrane proteins and on being able to follow their behavior and their interactions with selected lipid components.

REFERENCES

ABRAM, D., and GIBBONS, N. E. (1961). Can. J. Microbiol. 7, 741.

ABRAMSON, M. B., KATZMAN, R., WILSON, C. E., and GREGOR, H. P. (1964). J. Biol. Chem. 239, 4066.

ABRAMSON, M. B., COLACICCO, G., CURCI, R. and RAPPORT, M. M. (1968). Biochemistry 7, 1692.

BAXTER, R. M. (1959). Can J. Microbiol. 5, 47.

BROWN, A. D. (1963). Biochim. Biophys. Acta 75, 425.

BROWN, A. D. (1965). J. Mol. Biol. 12, 491.

BROWN, A. D., and PEARCE, R. F. (1969). Can. J. Biochem. 47, 833.

CHAPMAN, D., and KAMAT, V. B. (1968). "Regulatory Functions of Biological Membranes." Elsevier, Amsterdam.

CHEN, J. S., BARTON, P. G., BROWN, D., and KATES, M. (1974). Biochim. Biophys. Acta 352, 202.

ESSER, A. F., and LANYI, J. K. (1973). Biochemistry 12, 1933.

JOST, P. C., GRIFFITH, O. H., CAPALDI, R. A., and VANDERKOOI, G. (1973). Proc. Nat. Acad. Sci. U.S.A. 70, 480.

KATES, M. (1972). In "Ether Lipids; Chemistry and Biology" (F. Snyder, ed.), p. 351. Academic Press, New York.

KATES, M., PALAMETA, B., and YENGOYAN, L. S. (1965). Biochemistry 4, 1595.

KATES, M., and HANCOCK, A. J. (1971). Biochim. Biophys. Acta 248, 254.

KELLY, M., NORGARD, S., and LIAANEN-JENSEN, S. (1970). Acta Chem. Scand. 24, 2169.

KRAMER, J. K. G., KUSHWAHA, S. C., and KATES, M. (1972). Biochim. Biophys. Acta 270, 103.

KUSHWAHA, S. C., PUGH, E. L., KRAMER, J. K. G., and KATES, M. (1972). Biochim. Biophys. Acta 260, 492.

KUSHWAHA, S. C., GOCHNAUER, M. B., KUSHNER, D. J., and
KATES, M. (1974). Can. J. Microbiol. 20, 241.

LANYI, J. K. (1969a). J. Biol. Chem. 244, 2864.

LANYI, J. K. (1969b). J. Biol. Chem. 244, 4168.

LANYI, J. K. (1971). J. Biol. Chem. 246, 4552.

LANYI, J. K. (1974a). Bacteriol. Revs. 38, 272.

LANYI, J. K. (1974b). Biochim. Biophys. Acta 356, 245.

LANYI, J. K., and STEVENSON, J. (1970). J. Biol. Chem. 245,
4074.

LANYI, J. K., PLACHY, W. Z., and KATES, M. (1974).
Biochemistry 13, 4914.

LARSEN, H. (1967). In "Advances in Microbial Physiology"
(A. H. Rose and J. F. Wilkinson, eds.) Vol. 1, 97,
Academic Press, New York.

LIEBERMAN, M. M., and LANYI, J. K. (1971). Biochim. Biophys.
Acta 245, 21.

MESCHER, M. F., STROMINGER, J. L., and WATSON, S. W. (1974).
J. Bacteriol. 120, 945.

ONISHI, H., and KUSHNER, D. J. (1966). J. Bacteriol. 91, 646.

PLACHY, W. Z., LANYI, J. K., and KATES, M. (1974).
Biochemistry 13, 4906.

REISTAD, R. (1970). Arch. Mikrobiol. 71, 353.

SINGER, S. J., and NICOLSON, G.L. (1972). Science 175, 720.

STOECKENIUS, W., and ROWEN, R. (1967). J. Cell Biol. 34, 365.

STOECKENIUS, W., and KUNAU, W. H. (1968). J. Cell Biol. 38,
337.

TORNABENE, T. G., KATES, M., GELPI, E., and ORO, J. (1969).
J. Lipid Res. 10, 294.

VISENTIN, L. P., CHOW, C., MATHESON, A. T., YAGUCHI, M., and
ROLLIN, F. (1972). Biochem. J. 130, 103.

MAGNETIC RESONANCE STUDY OF ION INTERACTION
WITH BACTERIAL SURFACES

Robbe C. Lyon, Nancy S. Magnuson, James A. Magnuson

INTRODUCTION

Magnetic resonances of ions of quadrupolar nuclei have
been used in many biochemical studies (Dwek, 1973; Jardetzky
and Wade-Jardetzky, 1971). Among the common nuclei studied
are ^{35}Cl (Magnuson and Magnuson, 1972; Stengle and
Baldeschwieler, 1967; Ward, 1969); ^{23}Na (James and Noggle,
1969); ^{81}Br (Collins et al., 1973); and ^{25}Mg (Magnuson and
Bothner-By, 1971). In the last several years the ^{23}Na
resonances have been used extensively to probe the state of
ions in biological tissue. This has led to some serious
problems in interpretation, as many of the spectra of these
ions are complicated by electric field gradients within the
sample.

Early studies on ^{23}Na resonances in muscle tissue detected
only 40% of the signal expected from the actual sodium content
(Hazelwood, 1973). This work was later explained by several
research groups. Shporer and Civan (1972) and Berendson and
Edzes (1973) have demonstrated clearly that the absence of
60% of the expected signal is related to the presence of the
sodium ions in anisotropic environments. We have examined
the interaction of sodium ions with several membrane surfaces
by nuclear magnetic resonance spectroscopy. Both line widths
or relaxation times, and integrated intensities can be used

to study the surface properties of cells.

The resonance of ^{23}Na ions in dispersions of bacteria are characterized by broadened lines and reduced signal intensities. The spectral changes can be correlated with the surface properties of several species of bacteria. Although the phenomena associated with the magnetic resonances of ions in these heterogeneous systems is not completely understood, we present evidence that the technique will be valuable in studying surface characteristics of membranes.

MATERIALS AND METHODS

Bacteria were harvested at late log phase by centrifugation. They were resuspended in the appropriate NaCl solution and recentrifuged. The pellet was then dispersed in the same NaCl solution. Bacteria concentrations were determined by dry weight and are expressed as milligrams per milliliter.

Continuous wave NMR spectra were obtained with a modified Varian DP-60. The V-4210 radiofrequency unit was locked to a General Radio GR-1164 frequency synthesizer which was swept with a voltage ramp supplied by a Nicolet 1072 Time Averaging Computer. The magnetic field was locked at 14,000 gauss by using an external water proton signal resonating at 60.0 MHz. Line widths were determined by direct reading of the radio frequency. Integrations were performed with the Time Averaging Computer. Errors in the values reported for both line widths and integration values are probably $\pm$ 5%. Sodium resonances were observed at 15.87 MHz and chlorine resonances were observed at 5.87 MHz.

RESULTS

Line Widths

Two spectra of ^{23}Na ions are shown in Figure 1. The line width of the sodium resonance in the bacteria dispersion is noticeably broadened with Aerobacter aerogenes present. This is noticeable with all bacteria, but is especially apparent with gram-negative species.

In a simplified approach to the problem, we assumed that the spectra observed represent contributions from all sodium species present. Under conditions of rapid exchange, one can then assume that the line width observed, $\Delta\nu_{obs}$, is a

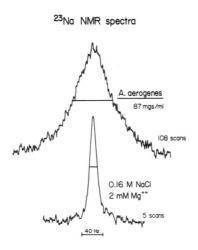

²³Na NMR spectra

Fig. 1. ^{23}*Na magnetic resonance spectra. The top spectrum is obtained from a dispersion of* A. aerogenes *in 0.16 M NaCl, 0.002 M MgCl$_2$. The lower spectrum is of 0.16 M NaCl, 0.002 M MgCl$_2$.*

weighted average of those representing all species.

$$\Delta\nu_{obs} = \Delta\nu_f X_f + \sum_i \Delta\nu_{bi} X_{bi}$$

The line widths of the free and bound species are $\Delta\nu_f$ and $\Delta\nu_b$, respectively, and the X's refer to respective fractions. It must be noted that $\Delta\nu_b$, the line width of the bound species, may not actually represent binding. As will be demonstrated below, this probably represents a line width produced when the ion experiences the field gradient from the bacterial surface. In effect, the observed line width is produced by the sum of many nuclei experiencing

307

different gradients at different distances from the bacterial surface. If X_f is close to 1.0, then $\Delta\nu_f X_f$ can be represented by the line width of the standard NaCl, $\Delta\nu_{std}$. For reasons which will become apparent, the fraction of free sodium decreases significantly from 1.0 at high bacteria concentrations. Thus, this simplified treatment is qualitatively valid only at low bacteria concentrations. Further, if one replaces the summation over all sites i by an average bound line width $\Delta\nu_{avg}$, then one obtains the following:

$$\Delta\nu_{obs} - \Delta\nu_{std} = \Delta\nu_{avg} X_b$$

As a measure of X_b, we have used the dry weight of bacteria in milligrams per milliliter. A plot of $\Delta\nu_{obs} - \Delta\nu_{std}$ versus dry weight of bacteria should then give a straight line passing through the origin. Figure 2 shows this plot for <u>Sarcina ureae</u>. We have examined this type of data for over 20 species of bacteria and a plot of some of these is shown in Figure 3. As will be shown in what follows, these plots do vary from linearity at low concentrations of bacteria. This clearly shows that the model assumed is not completely accurate.

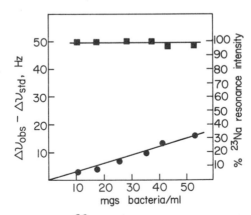

Fig. 2. Line width of ^{23}Na resonances as a function of <u>S. ureae</u> concentration, ●. This is plotted according to the simplified approach given in text. The percentage of maximum expected intensity is also plotted as a function of bacteria concentration, ■.

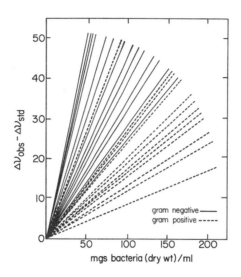

Fig. 3. A plot, as in Figure 2, for 26 bacteria species.

It had previously been demonstrated that the interaction
of sodium ions with both red blood cells and Acholeplasma
laidlawii membranes could be detected with [23]Na resonances,
and that these interactions could be influenced by the state
of the membrane and the nature of other metal ions competing
for negative sites on the membrane (Magnuson and Magnuson,
1973a). To investigate the surface of bacteria we examined
line width changes as a function of pH and also added metal
ion. The titration curve for A. aerogenes is shown in
Figure 4. Also shown in this figure is the titration curve
for the lipopolysaccharide (LPS) component prepared from the
cell envelope of A. aerogenes (Weinbaum *et al.*, 1971). The
titration curves are nearly identical, and the main
contributing component to the negative surface may be the
LPS. Gram-positive species do not contain LPS, but have other
polymers in their envelopes. Upon adding divalent metal ions,
such as $ZnCl_2$, $MgCl_2$, and $CaCl_2$, the Na resonance narrows. With
gram-negative species and the purified LPS, the three divalent
ions are equally effective. Figure 5 presents competition
curves for whole bacteria. With gram-positive species, however,

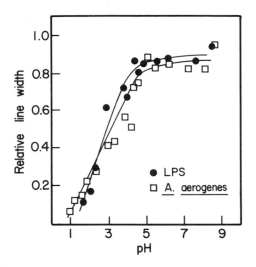

Fig. 4. The ^{23}Na line width as a function of pH for dispersions
of A. aerogenes and lipopolysaccharide (LPS) from
A. aerogenes. The line widths have been normalized
to 1.0 at pH 7.0, and this relative line width is
plotted versus pH.

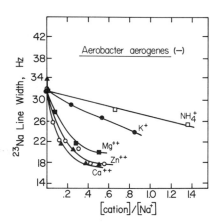

Fig. 5. Reduction in line width is shown as a function of
increasing cation concentration for dispersions of
A. aerogenes. Additions of cation, as the chloride
salt, were made to dispersions of bacteria in 0.16 M
NaCl.

Ca^{2+} appears to be more effective than Mg^{2+} which, in turn, is more effective than Zn^{2+} in shielding the bacteria from sodium ions. The three divalent ions shield LPS to a similar extent. This again implies that the LPS is contributing greatly to the surface charge properties of the gram-negative species. Similar studies with teichoic acid, a negatively charged component of gram-positive species (Baddiley, 1972), have not been successful.

The line widths do qualitatively reflect several properties of the cell surface. The relaxation phenomena are complex, however, and actual measurements of the spin-lattice relaxation times will be necessary. In addition, a more complex model is needed. To obtain a better idea of the relaxation processes, we undertook a study of the intensities of the sodium resonances.

Intensities of Resonances

We have previously reported that signal intensities, per cent of that expected from total sodium present, vary from 40-100% in dispersions of bacteria (Magnuson and Magnuson, 1973b). Berendson and Edzes (1973) have presented a rather complete picture of the phenomenon. Many of the dispersions of gram-negative bacteria give signals much lower than that expected for 100% of the sodium ions. The gram-positive species usually give signals close to 100% for similar concentrations. This extent of reduction depends on the species of bacteria employed. Figure 6 presents the data for two samples of Pseudomonas aeruginosa, a gram-negative bacterium. With 0.01 M Mg^{2+}, more bacteria are required to reduce that signal to 40%. This general behavior is also reflected in the line widths that are presented in Figure 7. When Mg^{2+} interacts with the surface, the sodium ion interaction, as reflected in both intensities and line widths, is reduced.

To examine the effect on intensities, a pH titration was carried out on the gram-negative A. aerogenes. The data for both line width and intensities are shown in Figure 8. As the pH is decreased, the surface is protonated. As the surface is reduced in negative charge, the line width is decreased showing less sodium ion interaction. The intensity is increased as the pH is lowered. Lower intensity corresponds to more interaction of the sodium with electric fields set up by the bacteria.

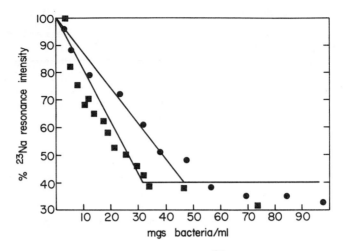

Fig. 6. *Integrated intensities for* ^{23}*Na resonances for* <u>Ps</u>.
<u>aeruginosa</u> *in 0.16 M NaCl and 0.01 M MgCl$_2$,●, and
in 0.16 M NaCl and 0.001 M MgCl$_2$,■.*

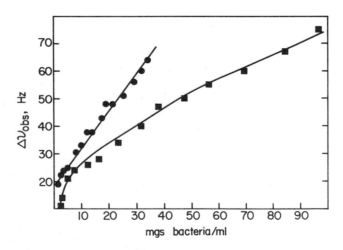

Fig. 7. *Line widths for* ^{23}*Na resonances for* <u>Ps</u>. <u>aeruginosa</u>
*in 0.16 M NaCl   and   0.01 M MgCl$_2$,■, and in 0.16 M
NaCl and 0.001 M MgCl$_2$,●.*

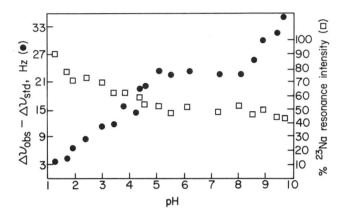

Fig. 8. Line widths and resonance intensities for a dispersion of A. aerogenes in 0.16 M NaCl as a function of pH.

Dispersions of Ps. aeruginosa were examined in varying concentrations of NaCl. Intensities were obtained for different sodium ion concentrations, whereas the bacteria concentration was held constant. These data are presented in Figure 9. The concentration of bacteria was high enough so that approximately 25% of the solution volume was taken up by bacteria. At lower salt concentration, the resonance intensity is markedly reduced. At lower salt concentrations, more of the total sodium is interacting with the bacterial surface.

In addition to examining ^{23}Na resonances in the bacterial dispersions, we obtained ^{35}Cl resonances in several samples. The ^{35}Cl intensities were essentially 100% of the maximum expected, even in samples demonstrating ^{23}Na signals of only 40%. The sodium ions obviously spend more time in the electric field gradient, or get closer to the bacterial surface, than do the chloride ions.

DISCUSSION

The line widths of the sodium resonances can be treated as being composed of contributions from several different sodium types. One is free sodium which is not influenced by

313

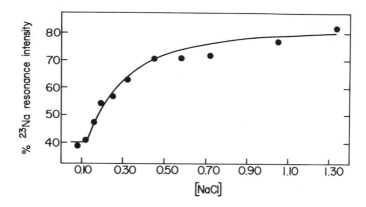

Fig. 9. *Integrated intensities for ^{23}Na resonances from a dispersion of Ps. aeruginosa. The NaCl concentration was varied, while the bacteria concentration was held constant at 30 mg/ml.*

the cells. Resonance contributions at high bacteria concentration from this sodium type will be negligible. Another contribution will be from sodium exchanging directly with the cell surface. This may be contributing to line widths at low bacteria concentration. In the intermediate regions two other sodium types probably predominate. These types are sodium ions which experience the field gradient from the bacterial surface. The other is from sodium within the cells. At higher concentrations of bacteria, this can be considerable.

Berendson and Edzes (1973) have thoroughly considered the effects of electric field gradients on the resonances of sodium ions. Any sodium ion experiencing a net field gradient greater than zero over a time of approximately 10^{-8} seconds should give a resonance split into three lines. In a heterogeneous sample, the outer two lines will vary in position because of a range of field gradients and orientations relative to the magnetic field. These satellites will normally not be visible under such conditions. It is these satellites which account for the missing resonance intensity in our samples. Not only do ions residing within the field exhibit only a 40% intensity line, but ions outside

of the field which diffuse into the electric field gradient
must also exhibit this kind of signal.

It is of interest, therefore, to examine both the
diffusion band around the bacteria and the band or layer of
potential drop around the bacteria. In a naive treatment,
one can assume that at a value of 40% intensity, all the ions
are experiencing an electric field gradient (approximately
40 mg/ml of bacteria in Figure 5). If one calculates the
distance between bacteria at this point, and one assumes the
sodium ions are evenly distributed through the solution, one
has to assume that the electric field gradient drops to zero
at a distance of 5000 Å from the bacterial surface. This is
necessary to allow all the sodium to experience the electric
field gradient.

Clearly, this 5000 Å is too large a distance, as all
calculations of surface potential require that it drop to
almost nothing within 50 Å (Verwey and Overbeek, 1948).
One must consider, however, the diffusion of ions into this
region of potential drop. Berendson and Edzes (1973) have
shown that the mean square displacement of a sodium ion in
10^{-8} seconds is less than 100 Å. Any ions within a 100 Å
band might diffuse into the electric field gradient
surrounding the bacteria. It seems likely that any sodium
ion experiencing this gradient must be within 150 Å of the
bacterial surface. This is a high estimate as the potential
will decrease to 1% of the surface potential at 30 Å. It is
of interest that the chlorine ions are relatively unaffected
by this field gradient. A spectrum giving only 50% of the
expected sodium resonance intensity gave a ^{35}Cl resonance with
90% of the expected chloride signal. The ^{35}Cl resonance
should have a reduced intensity just like the sodium resonance.
Obviously the sodium ions are adsorbed more to the surface than
the chloride ions.

Based on calculations of the surface potential and values
at discrete distances, we are forced to conclude that most of
the sodium ions are confined to a rather small volume adjacent
to the bacterial surface. This predicts that rather high
concentrations of sodium ions must exist near the surface.
This is consistent with calculations from the Boltzmann
equation (McLaughlin et al., 1971), if one assumes a surface
potential of approximately - 300 mV. This potential can be
derived from an equation relating surface potential to
isoelectric points as given by Sherbet and coworkers (1972).
We used an isoelectric point of 2.17 for Ps. aeruginosa

(Harden and Harris, 1953). The high concentration of sodium ions at the surface is consistent with other models, such as that proposed by Gillespie (1970).

It is of interest to calculate a value of surface charge density based on the magnetic resonance data for sodium. The concentration of divalent ions C^{2+} which will be as effective in screening the surface potential as a concentration of monovalent ions C^+ is given by the following relation:

$$C^{2+} = \frac{(C^+)^2}{(272\sigma)^2}$$

The value of σ is the surface charge in electronic charges per square angstrom. This relation has been presented by McLaughlin *et al.*, (1971). We have applied this relation in two ways. From Figure 5 it can be determined that a concentration of 0.032 M Mg^{2+} reduces the sodium interaction by 50%. The line width is reduced half the way to the standard line width at 0.032 Mg^{2+} . Solving for σ from these values gives 0.33 charges/100 $Å^2$. An alternative method is to use integration values and the data presented in Figure 6. From the integration values one can determine the percentage of ions free from the field gradient and also the percentage experiencing the gradient. As an example, consider the case where a 70% integration value is observed. This represents 70% of the total maximum intensity that might be detected. This signal is composed of two parts. The sodium ions which do not experience an anisotropic field contribute 5/7 of the observed signal or 50% of the expected total maximum intensity. Ions in an anisotropic environment contribute 2/7 of the observed resonance or 20% of the maximum expected intensity. This represents an example where half of the total ions experience a field gradient and half of the ions are free in solution.

Precentage observed = Fraction in isotropic environment
× 100% + Fraction in anisotropic
environment × 40%.

From the observed signal intensity, one can determine the location of the ions according to our proposed model. By comparing the intensities in Figure 6, one can determine how much sodium is displaced from the anisotropic environment when the extra Mg^{2+} is present. We shall assume that the sodium displaced is equal to the added 0.009 M Mg^{2+} for a given bacteria concentration. Using this calculation for several values of bacteria concentrations, an approximate value of σ can be determined from the above relation. The σ so determined is 0.15 charges/100 $Å^2$.

These values of σ compare favorably to other determinations. Haydon (1961) has used electrophoresis to determine a charge density of 0.1 charges/100 $Å^2$ for Escherichia coli. Other higher cell types give similar charge densities (Sherbet et al., 1972).

The effect of bacterial concentration on the observed sodium resonance intensity is illustrated in Figure 6 for two samples of Ps. aeruginosa in 0.16 M NaCl. The two samples differ in magnesium concentration, 0.001 M Mg^{2+} and 0.01 M Mg^{2+}. The observed signal intensity is reduced to 40% at values of 32 mg/ml for the lower Mg^{2+} concentration and 47 mg/ml for the higher Mg^{2+} concentration. At these values it appears that nearly all of the sodium ions experience an anisotropic environment. Most of these ions would be experiencing an electric field gradient due to the cell surface, and a small fraction would be located within the cell. As the bacterial concentration is decreased, contributions from free ions to the observed intensity increase. A theoretical curve was calculated for each sample, as shown by the lines in Figure 6. Three different environments for the sodium ions were considered: within the cell, near the cell surface and free in solution. It was assumed that each of the bacteria requires a particular number of sodium ions associated with the cell surface. This number was calculated to be 9.6×10^7 and 5.0×10^7 for the 0.001 M Mg^{2+} and the 0.01 M Mg^{2+} samples respectively. As the bacteria density increased in a constant sodium concentration the number of sodium ions associated with bacterial cell surfaces increased, becoming a larger fraction of the total number of sodium ions. As the bacterial density increased, the fraction of sodium ions within the cells also increased. Both of these effects cause the fraction of sodium ions in anisotropic environments to increase, resulting in a reduction in the observed signal intensity. The fraction of free ions continually declines as the bacterial density is increased. The result of these contributions are illustrated by the calculated straight lines which decrease from 100-40% signal intensity in Figure 6. After the 40% signal intensity is reached, it should be maintained at higher bacterial densities, unless the actual volume of bacterial components excludes solution.

Our explanation of the magnetic resonance data appears to be consistent if the resonances are treated as contributions from sodium ions in several types of environments. The continuous line in Figure 9 is calculated by assuming that the bacteria make up 25% of the total volume. We assume that 25% of the sodium chloride solution is within the bacteria, and that this sodium is entirely within an anisotropic environment. A signal representing 40% of this fraction will be observed. At a value of 0.12 M NaCl, the intensity observed is 40%. At this point most of the ions are adsorbed near the surface in the anisotropic environment. Very little of the sodium is free. Although variations of 0.01 M make little difference in the calculations, we used 0.11 M NaCl as the portion within the adsorbed layers. This value was held constant throughout. The third parameter is the free sodium which is determined by subtracting the adsorbed sodium and internal sodium from the total sodium. The calculated line and experimental points are in close agreement.

By using resonance intensities as described in this manuscript, we have been able to sort out several contributions to the resonance phenomena. A thorough understanding is necessary in order for one to apply magnetic resonance techniques to heterogeneous samples. Heterogeneity has produced severe problems in interpretation in the past. Integration values reported herein most certainly reflect a more accurate picture of the experiment than does the qualitative treatment of line widths. We have not yet defined all the factors influencing line widths. Work is now in progress which will measure T_1 values instead of line widths. With the work carried out to date, our treatment appears to be good. Even the line widths are in good qualitative agreement.

The magnetic resonance technique allows one to study properties of the surface. Gram-negative bacteria interact more strongly than gram-positive bacteria, as demonstrated by both line widths and integration values. This correlates somewhat with structure of the cell envelope components. By examining different species or a given species grown under different conditions, these correlations should be established more conclusively. Similar magnetic resonance phenomena can be observed with dispersions of red blood cells and lipid vesicles. Extension of this treatment to systems of this type should yield information about surface charge, the resulting interaction with solution components, and the effects of compounds which alter the membrane state.

ACKNOWLEDGEMENTS

This research was supported in part by a grant from the Washington State University Research Committee, and in part by United States Public Health Service Grant CA 14496-02.

REFERENCES

BADDILEY, J. (1972) Essays in Biochemistry 8, 35.

BERENDSON, H.J.C., and EDZES, H.T. (1973) Ann. N.Y. Acad. Sci. 204, 459.

COLLINS, T.R., STARCUK, Z., BURR, A.H., and WELLS, E.J. (1973) J. Amer. Chem. Soc. 95, 1649.

DWEK, R.A. (1973) In "Nuclear Magnetic Resonance in Biochemistry", Chapter 13, Clarendon Press, Oxford.

GILLESPIE, C.J. (1970) Biochim. Biophys. Acta, 203, 47.

HARDEN, V.P., and HARRIS, J.O. (1953) J. Bact. 65, 198.

HAYDON, D.A. (1961) Biochim. Biophys. Acta, 50, 457.

HAZELWOOD, C.F. (ed.) (1973) "Physicochemical State of Ions and Water in Living Tissues and Model Systems", Ann. N.Y. Acad. Sci., 204, N.Y. Academy of Science.

JAMES, T.L., and NOGGLE, J.H. (1969) Proc. Nat. Acad. Sci. U.S. 62, 644.

JARDETZKY, O., and WADE-JARDETZKY, N.G. (1971) Ann. Rev. Biochem. 40, 605.

MAGNUSON, J.A., and BOTHNER-BY, A.A. (1971) In "Magnetic Resonances in Biological Research" (C. Franconi, ed.), 365, Gordon and Beach, London.

MAGNUSON, J.A., and MAGNUSON, N.S. (1972) J. Amer. Chem. Soc. 94, 5461.

MAGNUSON, J.A., and MAGNUSON, N.S. (1973a) Ann. N.Y. Acad. Sci. 204, 297.

MAGNUSON, N.S., and MAGNUSON, J.A. (1973b) Biophys. J. 13, 1117.

MCLAUGHLIN, S.G.A., SZABO, G., and EISENMAN, G. (1971) J. Gen. Physiol. 58, 667.

SHERBET, G.V., SLAKSHMI, M., and RAO, K.V. (1972) Exptl. Cell. Res. 70, 113.

SHPORER, M., and CIVAN, M.M. (1972) Biophys. J. 12, 114.

STENGLE, T.R., and BALDESCHWIELER, J.D. (1967) J. Amer. Chem. Soc. 89, 3045.

VERWEY, E.J.W., and OVERBEEK, J. Th. G. (1948) In "Theory of the Stability of Lypophobic Colloids", 25, Elsevier, New York.

WARD, R.L. (1969) Biochemistry 8, 1879.

WEINBAUM, G., KADIS, S. and AJL, S.J. (eds.) (1971) "Microbial Toxins", Vol. IV. Academic Press, New York.

THE INFLUENCE OF DIPOLE POTENTIALS ON THE MAGNITUDE AND THE KINETICS OF ION TRANSPORT IN LIPID BILAYER MEMBRANES

Gabor Szabo

INTRODUCTION

The main features of ion transport through lipid bilayer membranes are relatively well understood for large hydrophobic ions which permeate the membrane directly (Ketterer *et al.*, 1971) and for small hydrophilic ions which permeate the membrane by combining with a hydrophobic carrier molecule (Szabo *et al.*, 1969; Läuger and Stark, 1970; for a review see Haydon and Hladky, 1972).

These apparently different transport mechanisms are basically similar in that only a single charged species, the hydrophobic ion or ion-carrier complex, permeates the membrane; for both mechanisms the permeant ion simply "dissolves" in the hydrocarbon-like membrane interior. For both mechanisms, furthermore, the concentration of the permeant species in the membrane interior is small compared to that in the aqueous phases or at the membrane surfaces. Therefore, in both cases transport occurs through the nonpolar membrane interior between reservoirs of the permeant species which are located at the membrane surfaces and/or in the aqueous phases. The main difference between the two permeation mechanisms lies in the identity of the permeant species; in direct transport it is simply a hydrophobic ion, whereas in carrier mediated transport it is an ion-carrier complex formed by a chemical reaction.

The lipid composition of the bilayer membrane plays a major role in modulating direct (Szabo, 1974) as well as carrier-mediated transport of ions (Szabo *et al.*, 1972). Conceptually, the membrane composition may alter ionic permeability in two ways: (1) by altering the amount of permeant species in the surface reservoirs or, (2) by altering its translocation rate across the membrane interior. The lipid polar head groups are expected to be most effective in controlling the size of the surface reservoirs by virtue of chemical interactions with the permeant ion, or by virtue of ionized as well as dipolar head groups creating an electrical potential difference across the membrane-solution interface. In contrast, the lipid hydrocarbon tails are expected to be most effective in controlling the translocation rate of the permeant species across the membrane interior by virtue of their dimensions, motility, and polarity.

Despite the fact that membrane composition plays an essential role in ion permeation processes, little is known about the detailed mechanisms by which lipid composition alters various steps in ion transport.

The effects of charged polar head groups on the steady-state permeability properties of carrier-induced ion transport have been interpreted in terms of a double-layer potential (McLaughlin *et al.*, 1970)., which is presumed to simply alter the concentration of permeant ion (or ion-carrier complex) at the membrane surface.

Compensation potentials of lipid monolayers spread at the air-water interface (Davies and Rideal, 1963) indicate that substitution of lipid polar head groups can result in large changes of the electrostatic potential across the interface even in the *absence* of ionized polar head groups. Such changes in the "dipole" potential presumably result from changes in the dipole moment and packing of nonionic groups present at the interface. Recently, such dipole potentials have been postulated (LeBlanc, 1969; Szabo *et al.*, 1972; Hladky and Haydon, 1973) and demonstrated to exist (Szabo, 1974) in at least some lipid bilayer membranes as well.

The objective of this paper is to assess the magnitude of dipole potential changes and to determine the way in which these alter the various steps in the permeation of bilayer membranes by a particularly simple lipophilic ion, tetraphenylborate. Because of the similarities between direct and carrier-mediated transport, the conclusions reached here are expected to be relevant toward the understanding of the influence of dipole potentials on carrier-induced transport as well.

THEORETICAL

A salient feature of lipid bilayer membranes is a thin (4 nm), nonpolar interior formed by apposing lipid hydrocarbon tails.[1] It acts as a barrier towards ionic movement, since large energies are required to transfer a charged species from the polar aqueous phase into the hydrocarbon-like membrane interior. For common electrolytes (such as Na, K, or Cl) which have small ionic size, the energies of transfer are unfavorable enough to result in a negligibly small membrane permeability. In contrast, for large organic cations, such as tetraphenylborate (TFB), tetraphenylphosphonium (TFP), dipropyloxadicarbocyanine (CC_5), and carbonylcyanide m-chlorophenyl hydazone (CCCP), the energy of transfer is apparently favorable enough to result in significantly large membrane permeability. For convenience, we shall refer to these preferentially permeant ions as "permions."

Because of a more favorable energy of transfer, the permion is expected to be the predominant species in the interior of a bilayer membrane which is formed in aqueous solutions also containing common electrolytes, such as Na or Cl. Although the permion leaves its counterion behind in the aqueous phase, the resulting charge separation does not give rise to a significant electrical potential in the membrane interior provided that the permion concentration within the extremely thin membrane interior is small ($< 10^{-5}M$) (Läuger and Neumcke, 1973).

---

[1] Large electrical capacitance ($0.3-1\mu F/cm^2$) observed typically for bilayer membranes, is consistent with this view (Haydon and Hladky, 1972).

A permion should experience three types of interactions within a membrane. Short-range interactions result from collisions with lipid hydrocarbon tails and may be viewed as frictional forces; the effect of these on permion movement can be taken into account by a diffusion coefficient, $D_{(x)}$, determined by the size of the permion and by the consistency of the membrane interior. Long-range interactions result from coulombic interactions with polar molecules present mainly in the aqueous phase and may be viewed as equilibrium forces; the effect of these on permion movement can be taken into account by a position-dependent free energy, $\mu_m^O(x)$ also characteristic of the permion and of the membrane interior (Läuger and Neumcke, 1973). Lastly, any electrical potential $\psi(x)$ that exists in the membrane interior would also affect the flow of permions and must therefore be taken into account as an energy term $zF\psi(x)$. There are two obvious sources for such internal potentials. First, oriented dipoles located near the membrane surface could create an important potential difference between the membrane boundary and the membrane interior. This "dipolar" potential ψ_m is expected to vary little in the membrane interior and to fall off only near the membrane boundaries since there are no strongly dipolar groups located in the membrane interior. Second, any potential difference ϕ applied across the membrane boundaries must superimpose a gradient on the internal potential. This gradient should be approximately linear $\phi(x/d)$ in the hydrocarbon-like membrane interior of thickness d (Läuger and Neumcke, 1973). The overall internal potential, then, is the sum of these two potentials:

$$\psi(x) = \psi_m + \frac{\phi}{d} x \ .$$ (1)

Conductance and Permeability of the Membrane Interior

Ionic current density J through the membrane can be described in terms of the previously defined interactions by the generalized Nernst-Planck (diffusion) equation (Läuger and Neumcke, 1973)

$$-J = zFD(x) \ c(x) \ \frac{d}{dx}\left[\frac{\mu_m^O(x)}{RT} + \frac{zF\psi(x)}{RT} + \ln \ c(x)\right]$$ (2)

where z is the valence of the permion, T is the temperature, R is the gas constant, and F the Faraday. Since there is only a single ion, the permion present in the membrane, the steady-state current through the membrane interior is easily obtained by integrating (2) across the membrane thickness.[1] Using Eq. (1) for symmetrical membranes, for which $\mu_m^o(x) = \mu_m^o(d - x)$ and $\psi_m(x) = \psi_m(d - x)$ the result is:

$$J = -zF \left[\int_0^{d/2} \frac{2 \cosh\left[\frac{zF\phi}{RT}(\frac{x}{d} - \frac{1}{2})\right]}{D(x) \exp\left[-\frac{\mu_m^o(x) + zF\psi_m}{RT}\right]} \right]^{-1}$$

$$\cdot \left[c(d) \exp(\frac{zF\phi}{2RT}) - c(0) \exp(-\frac{zF\phi}{2RT}) \right] \tag{3}$$

where $c(0)$ and $c(d)$ are respectively the concentrations of the permion at the left and right boundaries, and ϕ is the potential difference across the boundaries. It is convenient to define

$$P_m(\phi) = \left[\int_0^{d/2} \frac{2 \cosh\left[\frac{zF\phi}{RT}(\frac{x}{d} - \frac{1}{2})\right]}{D(x) \exp\left[-\frac{\mu_m^o(x) + zF\psi_m}{RT}\right]} \right]^{-1}. \tag{4}$$

All the information on membrane properties obtainable from electrochemical measurements is contained in $P_m(\phi)$. For small values of the applied potential ($\phi < 25$ mV) $P_m(\phi)$ reaches a well defined limit which is independent of ϕ.

---

[1] *The thinness of the membrane and the low ionic concentrations in the membrane imply that the current through the membrane interior should reach steady-state so quickly that under the time scale of experiments (milliseconds) the system is at steady-state. However, the boundary-concentrations may, and sometimes do vary on these time scales, giving rise to an observed current which decays with time.*

$$P_m = \lim_{\phi \to 0} P_m(\phi) = \left[2 \int_0^{d/2} \frac{dx}{D(x) \, \exp\left[-\dfrac{\mu_m^o(x) + zF\psi_m}{RT}\right]} \right]^{-1} . \quad (5)$$

P_m is the *permeability* of the membrane interior.[1] This can be seen by setting $\phi = 0$ in Eq. (3):

$$P_m = \frac{J/zF}{c(0) - c(d)} . \quad (6)$$

The zero-current *conductance* of the membrane interior G is proportional to P_m for membranes with identical boundary concentrations, $c(0) = c(d) = c_m$.

$$G = \lim_{\phi \to 0} \frac{-J}{\phi} = \frac{z^2 F^2}{RT} P_m c_m . \quad (7)$$

Equation (7) shows that, in principle, P_m can be deduced from conductance as well as from permeability measurements. However, the boundary concentrations $c(0)$ and $c(d)$, which are not directly available, must first be determined in order to do this.

Overall Membrane Conductance and Membrane Permeability

The results of Eqs. (3), (6), or (7) are directly useful if the boundary concentrations $c(0)$ and $c(d)$ are expressed as a function of the known bulk aqueous concentrations of the permion, c' and c'' respectively. Under the simplest conditions[2] equilibrium between the bulk aqueous phase and the membrane surface is not significantly perturbed by the flow

---

[1] *P_m has the appropriate units of centimeters per second when the concentrations are expressed in moles per cubic centimeter.*

[2] *A necessary but not sufficient condition is that the membrane permeability P be much smaller than that of the aqueous unstirred layers P_w (Läuger and Neumcke, 1973).*

of electric current. Whenever this condition is fulfilled a simple partition coefficient can be used to relate the surface concentrations to those in the bulk phases:

$$c(0) = \gamma c' \quad \text{and} \quad c(d) = \gamma c'' \tag{8}$$

where

$$\gamma = \exp\left(-\frac{\mu_s^o + zF\psi_s}{RT}\right) \tag{9}$$

is the adsorption coefficient of the permion, expressed in terms of the free energy difference μ_s^o and the potential difference ψ_s between the membrane *surface* and the *bulk* aqueous phase. For membranes bathed in solutions of identical composition, $c' = c'' = c$ so that $c(0) = c(d) = \gamma c$, the zero-current membrane conductance is obtained directly from Eq. (7)

$$G = \frac{z^2 F^2}{RT} P_m \gamma c . \tag{10}$$

The overall membrane permeability P can be calculated from the membrane conductance using Eq. (7)

$$P = P_m \gamma = \frac{RT}{z^2 F^2} \frac{G}{c} . \tag{11}$$

Clearly, P_m and γ cannot be determined separately from conductance measurements alone.[1] It is possible, however, to identify changes in the overall *chemical* and *electrical* potentials in the membrane by comparing permeabilities for suitable lipophilic catic ns and anions (Szabo, 1974). This is seen by noting that Eq. (5) can be approximated as

---

[1] It is necessary to carry out relaxation studies to separate P_m and γ. These are described in the next section.

$$P_m \simeq \frac{\exp[-zF\psi_m/RT]}{2\int_0^{d/2} \frac{dx}{D(x)\ \exp[-\mu_m^O(x)/RT]}} \qquad (12)$$

since ψ_m is approximately constant near the middle section of the membrane $(x \simeq d/2)$ which, due to maximal values of $\mu_m^O(x)$, practically determines the value of the integral in the denominator of Eq. (5) (Hall *et al.*, 1973). Inserting approximation (12) and definition (9) into Eq. (11) the desired separation of chemical and electrical potentials is obtained:

$$P = \frac{\exp[-\mu^O s/RT]}{2\int_0^{d/2} \frac{dx}{D(x)\ \exp[-\mu_m^O(x)/RT]}}\ \exp\left(-\frac{zF\psi_0}{RT}\right) \qquad (13)$$

where $\psi_0 = \psi_m + \psi_s$ is the potential difference between the membrane *interior* and the aqueous phase.[1] For simplicity, it is convenient to introduce the definition

$$\lambda = \frac{\exp[-\mu^O s/RT]}{2\int_0^{d/2} \frac{dx}{D(x)\ \exp[-\mu_m^O(x)/RT]}} \ . \qquad (14)$$

Equation (13) now simplifies to

$$P = \lambda\ \exp\left(-\frac{zF\psi_0}{RT}\right) \ . \qquad (15)$$

Note that λ is the membrane permeability that one would observe in the absence of any potential difference between

---

[1] *The following section shows that it is possible to separate* ψ_m *and* ψ_s *by performing measurements on the kinetics of transport.*

the membrane interior and the bulk aqueous phases. Therefore λ can be viewed as the "intrinsic permeability" of the
membrane. The Results section shows that for certain hydrophobic cations and anions intrinsic permeability is altered
in the same way by altering membrane composition. Under
these conditions, changes in λ and ψ_0 can be separated by
comparing permeabilities for cationic $(z = 1)$ and anionic
$(z = -1)$ species.

Kinetics of Transport

For permeant ions which adsorb strongly at the membrane
surface, it is convenient to introduce surface concentrations,
N' and N''

$$N' = \delta c(0); \quad N'' = \delta c(d) \tag{16}$$

in which δ is the thickness of the adsorption layers. Whenever a membrane is at equilibrium with aqueous solutions of
equal permion concentrations, c, the surface concentration N'
and N'' can be related to c by the simple partition relation:

$$N' + N'' = N_T = 2\gamma\delta c . \tag{17}$$

In the case of ions for which the membrane permeability is
much larger than that of the unstirred layers $(P \gg P_w$, see
Läuger and Neumcke, 1973) the adsorbed permion is effectively
trapped at the membrane surfaces by the unstirred layers.[1]
Therefore, the total *number* of charges at the membrane surface, N_T, will remain approximately constant so that

$$N' + N'' = N_T . \tag{18}$$

When permions are transferred across the membrane by an
applied potential, N' and N'' will change, but in such a way

---

[1]*In the case of ions for which $P \ll P_w$, equilibrium will
be maintained by diffusion from and into the unstirred layers
so that the surface concentrations, and therefore the membrane current, will show no time dependence.*

that Eq. (18) is obeyed since transport through the unstirred layers is negligibly small. Consequently, the ionic current through the membrane will decay, *not* as a result of a time-dependent conductance of the membrane interior, but simply because of the altered boundary concentrations.

The transmembrane current can be written in terms of surface concentrations by introducing Eqs. (16) and (5) into Eq. (4):

$$J(t) = -zF \frac{P_m(\phi)}{\delta} \left[N'' \exp\left(\frac{zF\phi}{2RT}\right) - N' \exp\left(-\frac{zF\phi}{2RT}\right) \right] . \quad (19)$$

It should be emphasized that $J(t)$ is a pseudo steady-state current in which conduction through the membrane interior is already at steady-state but the boundary concentrations still vary (De Levie *et al.*, 1974). The parameter P_m/δ has units of reciprocal seconds. It can therefore be viewed as a rate constant for transfer across the membrane interior (Ketterer *et al.*, 1971).

Since the permeant ion is trapped at the membrane surfaces, the rate with which its concentration changes at each surface is equal to the number of ions that are translocated:

$$-\frac{dN'}{dt} = \frac{dN''}{dt} = \frac{J(t)}{zF} . \quad (20)$$

Therefore, the surface concentrations N' and N'' must obey the differential equation

$$\frac{dN'}{dt} = \frac{P_m}{\delta} \left[N'' \exp\left(\frac{zF\phi}{2RT}\right) - N' \exp\left(-\frac{zF\phi}{2RT}\right) \right] . \quad (21)$$

In the "voltage clamp" experiments carried out to characterize the kinetics of transport, the membrane potential was held at some small value ϕ and then at $t = 0$ changed to $-\phi$. The pseudo-steady-state ionic current is obtained directly for these initial conditions by solving Eq. (21). The result is:[1]

$$J(t) = J(0) \, e^{-t/\tau} \quad (22)$$

where

$$J(0) = -2zF \, N_T \, \frac{P_m}{\delta} \, \sinh\left(\frac{zF\phi}{2RT}\right) \tag{23}$$

and

$$\frac{1}{\tau} = \frac{2P_m}{\delta} \, \cosh\left(\frac{zF\phi}{2RT}\right) . \tag{24}$$

These results are formally equivalent to those derived by Ketterer *et al.* (1971) and reviewed by others (Haydon and Hladky, 1972). In the experimentally relevant limit of small applied potentials and using Eq. (17) to express the value of N_T, Eqs. (23) and (24) reduce to

$$J(0) = -2 \, \frac{z^2 F^2}{RT} \, \gamma P_m \, c\phi \tag{25}$$

and

$$\frac{1}{\tau} = \frac{2P_m}{\delta} . \tag{26}$$

The initial value of the membrane conductance

$$G(0) = \frac{-J(0)}{2\phi} = \frac{z^2 F^2}{RT} \, P_m \gamma c \tag{27}$$

is that expected for equilibrium of surface adsorption, see Eq. (10). Thus, the initial current simply measures the membrane permeability:

$$P = \frac{J(0)RT}{2\phi cz^2 F^2} = \frac{G(0)}{c} \, \frac{RT}{z^2 F^2} = P_m \gamma . \tag{28}$$

---

[1]For perfectly trapped permions, the steady-state current should vanish. However, as discussed by Haydon and Hladky (1972), the finite permeability of the unstirred layers permits a small flow of the permion through the unstirred layers.

In contrast, the inverse of the relaxation time, $1/\tau = 2P_m/\delta$ yields within a constant, the permeability of the membrane interior alone. Since

$$\frac{P_m}{\delta} = \frac{1}{2\tau} \tag{29}$$

and

$$\gamma\delta = 2\tau P \tag{30}$$

it is possible to deduce separately the way in which membrane composition alters translocation (P_m/δ) across the membrane interior, or adsorption ($\gamma\delta$) onto the membrane surface.

EXPERIMENTAL METHODS

Permeabilities for four hydrophobic ions were determined in lipid bilayer membranes formed from various mixtures of cholesterol and monoolein. The membranes were formed following well established techniques (Szabo et al., 1969) on a Teflon partition separating two small aqueous reservoirs (either 20 ml or 4 ml each) containing aqueous solutions of identical composition. Monoolein (Sigma grade, 99%) and cholesterol (Eastman, reagent grade) were dissolved in n-decane (Eastman, practical grade) so that the total lipid concentration was 25 mM but the cholesterol mole fraction, X_{CHOL}, varied from 0 to 0.91. Bilayer membranes formed from these lipids were stable (life time over 15 min). No reliably stable membranes could be formed from lipids with $X_{CHOL} > 0.91$. The lipids were used without further purification. Experiments using highly purified cholesterol (Applied Science, 99+%), or different batches of monoolein, gave identical results. The four hydrophobic ions, tetraphenylborate, TFB; tetraphenylphosphonium, TFP, (both J.T. Baker reagent grade); carbonylcyanide-m-chlorophenylhydrazone, CCCP, (Calbiochem); and 3,3'-dipropyloxydicarbocyamine iodide, CC_5 (a gift from Dr. Allan Waggoner), were used without further purification. The concentration of hydrophobic ions was increased in the aqueous solutions bathing the membrane by adding small aliquots of a stock solution. These were prepared by dissolving TFB or TFP in water or by dissolving CCCP or CC_5 in ethanol. Small amounts of ethanol have no effect on bilayer properties (Szabo et al., 1969).

Membrane permeability was characterized by recording the electrical current which resulted from a square pulse of potential applied across the membrane. A Keithley 427 current amplifier was used to measure these relatively small membrane currents. The output of the current amplifier was recorded either on a strip-chart recorder (for time independent currents) or on Polaroid film from an oscilloscope trace (for decaying membrane currents). Large chlorided silver plates were used to apply the potential which alternated slowly (typically 0.5 Hz) between small values (typically, ±10 mV). A separate pair of chlorided silver electrodes was used to verify the absence of electrode polarization. In order to prevent any voltage drop across the aqueous phases, 1 M NaCl solutions were used as an indifferent electrolyte for CC_5 and TFP. The solutions also contained $10^{-3}M$ phosphate (pH 6.0) for TFB and CCCP. Buffering was necessary for CCCP since it is a weak acid.

For TFP, CCCP, and CC_5 the currents were time-independent beyond the decay of the capacitive current (usually beyond .1 msec). Membrane conductance was obtained in these cases by dividing the membrane current by the applied potential. Conductance values were normalized to 1 cm^2 of membrane area.

For TFB the membrane current was found to decay with time. The exponential time course was analyzed according to the procedures described in the Results section. Initial values of the membrane conductance were determined from the membrane current extrapolated to zero time on a logarithmic plot. These values were close to those measured immediately following the decay of the capacitive current.

RESULTS

Assessment of Dipolar Surface Potential Changes in Monoolein-cholesterol Bilayers

Two permeant cations (TFP, CC_5) and two permeant anions (TFB, CCCP) were used to deduce the sign and magnitude of the dipole potential changes in monoolein membranes of increasing cholesterol content. The chemical structures of these ions are shown in Figure 1. Note that TFP and TFB have nearly identical size, shape, and chemical structures. The main difference between these two ions is in their electric charge: TFP is a cation ($z = 1$) and TFB is an anion ($z = -1$). In contrast, the size, shape, and structure of CC_5, which is a large cation, and that of CCCP, which is a large anion, are different from each other and also from those of TFP and TFB.

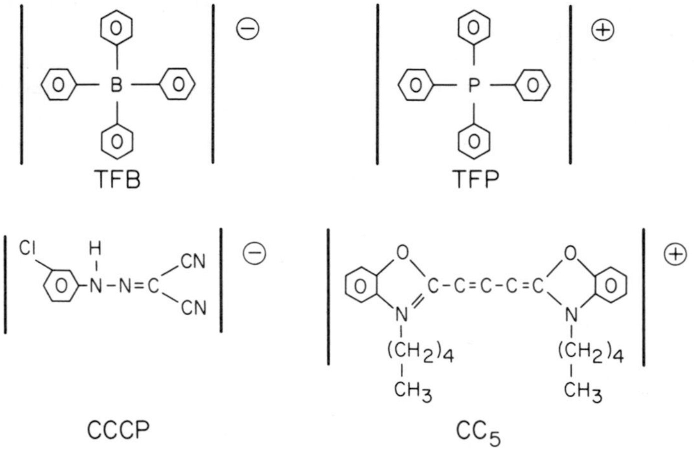

Fig. 1. Chemical structures for the four hydrophobic ions
that were used to determine surface potential
changes in bilayer membranes. TFB: tetraphenylborate; TFP: tetraphenylphosphonium; CCCP: carbonylcyanide m-chlorophenylhydrazone; CC_5: 3,3'-dipropyl-
oxadicarbocyanine.

Figure 2 shows, on a logarithmic scale, the permeability
P for these four permions in monoolein membranes as a function of increasing cholesterol mole-fraction, X_{CHOL}. The
permeabilities were calculated, using Eq. (11), from membrane
conductances measured at small potentials.[1] The most prominent feature in Figure 2 is the opposite effect that increasing X_{CHOL} has on the permeability for negative and positive

---

[1]For TFP, CC_5, and CCCP, these conductances were time
invariant. For TFB, for which the conductance decayed with
time, P was calculated from the initial value of the conductance (see Eq. (28)).

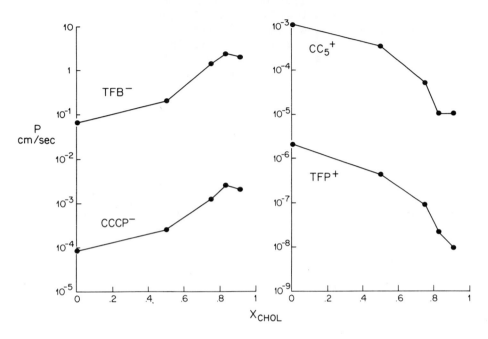

Fig. 2. *The opposite effect of increasing cholesterol content on cationic (left panel) and anionic (right
panel) permeability in monoolein bilayers. X_{CHOL} is
the mole fraction of cholesterol in the lipid solution (25 mM in decane) from which the membranes were
formed. Absolute values of P's for CCCP are not
meaningful since CCCP is not completely ionized at
pH 6.*

ions. For the two negative ions (left panel) increasing
X_{CHOL} has the effect of *increasing* P in the same way, despite
the fact that TFB and CCCP have rather different molecular
structures. For the two positive ions in contrast (right
panel), increasing X_{CHOL} has the effect of *decreasing* P in
the same way, also despite the fact that TFP and CC_5 have
rather different molecular structures. Clearly, it is not
the details of molecular structure, size or shape, but the
charge that is the relevant variable. The potential in the
membrane appears to become more positive as X_{CHOL} increases.

According to Eq. (15) the permeability changes that are seen in Figure 2 result either from changes in intrinsic conductance, λ, or changes in surface potential, ψ_0. Since absolute values of λ or ψ_0 cannot be determined separately, it is convenient to rewrite Eq. (15) in terms of variables referred to a pure monoolein membrane, identified here by the superscript MO.

$$\bar{P} = \frac{P}{P^{MO}} = \frac{\lambda}{\lambda^{MO}} \exp \frac{zF(\psi_0 - \psi_0^{MO})}{RT} . \tag{31}$$

The exponential factor in Eq. (31) should be the same for ions having the same charge. Furthermore, ratios of $\bar{P}$'s for CCCP and TFB (1, 1.07, 0.71, .89, .81) as well as those for CC_5 and TFP (1, .72, .90, 1.1, .47) are close to unity for $X_{CHOL} = 0, 0.5, 0.75, 0.83, 0.91$ indicating, to a first approximation, that λ/λ^{MO} has the same value for ions of the same charge. TFB and TFP are identical species except for their charge so that λ/λ^{MO} must also have the same value for ions of the opposite charge. Thus, to a first approximation, $\Delta\psi_0 = \psi_0 - \psi_0^{MO}$ and λ/λ^{MO} depend only on the membrane structure and not on the chemical properties of the permeant ions. Under these conditions Eq. (31) becomes for cationic ($\bar{P}_+$) and anionic ($\bar{P}_-$) permeabilities

$$\bar{P}_+ = \frac{\lambda}{\lambda^{MO}} \exp(-\frac{F\Delta\psi_0}{RT}) \qquad \bar{P}_- = \frac{\lambda}{\lambda_{MO}} \exp(\frac{F\Delta\psi_0}{RT}) . \tag{32}$$

From the ratio and the product of $\bar{P}$'s, $\Delta\psi_0$ and λ/λ^{MO} can be evaluated separately

$$\Delta\psi_0 = \frac{RT}{2F} \ln(\frac{\bar{P}_+}{\bar{P}_-}) \qquad \frac{\lambda}{\lambda^{MO}} = (\bar{P}_+ \bar{P}_-)^{1/2} . \tag{33}$$

Figure 3 plots as a function of X_{CHOL} the values of $\Delta\psi_0$ and $\log \lambda/\lambda^{MO}$ calculated from the data of Figure 2 using Eq. (33) in which $\bar{P}$'s averaged for TFP and CC_5 were used as $\bar{P}_+$ and $\bar{P}$'s averaged for TFB and CCCP were used as $\bar{P}_-$. The upper part of Figure 3 shows that the electrical potential inside the membrane becomes more positive as the membrane cholesterol content increases. This increase in ψ_0 is large; it appears to reach a limiting value of ~100 mV for predominantly cholesterol membranes. In contrast, λ/λ^{MO} decreases by

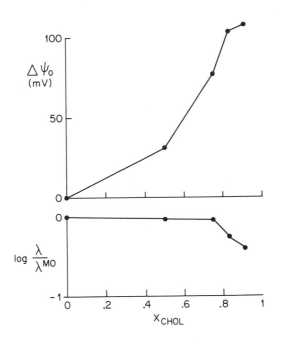

Fig. 3. *The effect of cholesterol on surface potential (upper scale) and relative permeability (lower scale) of monoolein bilayers.*

only a small factor, indicating--Eq. (14)--that the chemical potential of partition into the membrane and/or the diffusion within the membrane interior are not altered much by increasing cholesterol content.

The most likely origin of surface potentials is in ionized or strongly dipolar lipid head groups (Davies and Rideal, 1963). Since there are no ionizable head groups in either monoolein, which only have carbonyl and hydroxyl residues, or cholesterol, which only has a hydroxyl residue, the $\Delta\psi_0$'s seen in Figure 3 must be attributed to composition dependent changes in the strength, orientation, or density of dipolar surface groups. The fact that within experimental error, the results of Figure 3 are independent of the aqueous ionic strength further verifies that the $\Delta\psi$'s do not arise from a net surface charge (Szabo *et al.*, 1972).

The results of Figure 3 establish the existence of surface potentials of dipolar origin and also show that such "dipole potentials" alter ionic permeability to a considerable extent. Kinetic measurements of the type described in the next section are necessary, however, to establish the detailed mechanisms by which dipole potentials alter various steps in the transport of ions.

The Influence of Dipole Potentials on the Kinetics of Tetraphenylborate Transport

For TFB (but not for TFP, CC_5, or CCCP) the membrane current decays following a step change in the membrane potential. Typical initial time course of the current is shown in Figure 4. The logarithm of the membrane current, plotted as

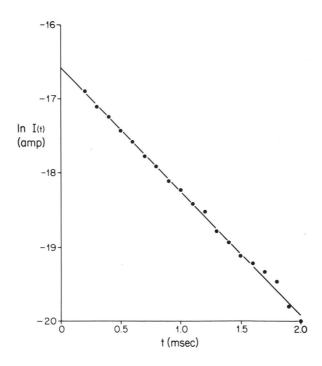

filled circles, is seen to decrease linearly with time. The regression line which connects the data points has a correlation coefficient $r = 0.999$ indicating that the current decays exponentially with time. An exponential decay ($r > 0.9$) of the early TFB current is seen for all membranes of different lipid composition, whereas at longer times the TFB current ceases to be exponential, and its decay becomes proportional to the inverse square root of time. The time course of relaxation observed here for cholesterol-monoolein membranes is qualitatively similar to that reported by LeBlanc (1969) and Ketterer et al. (1971) for phospholipid membranes.

Time dependence of the transmembrane TFB current is not unexpected since the membrane permeability P deduced from zero-time intercepts of the membrane current using Eq. (11), is always much greater ($P > .1$ cm/sec, see Figure 2) than that of the aqueous unstirred layers, P_w ($P_w \simeq 5 \times 10^{-6} \div 0.02$ $= 2.5 \times 10^{-4}$ cm/sec for aqueous diffusion coefficient of TFB of 5×10^{-6} cm^2/sec and an unstirred layer thickness of $2 \times .01$ cm). However, as discussed by Haydon and Hladky (1972) and by Läuger and Neumcke (1973), the decay of the membrane current is not expected to be exponential if the only cause of the time dependence is polarization of the aqueous unstirred layers. Furthermore, the experimentally observed TFB currents in monoolein membranes (as well as in phospholipid membranes) are larger, by orders of magnitude, than those expected for unstirred layers alone[1] [see

---

Fig. 4.   Typical time course of the TFB current. The natural logarithm of the membrane current, ln I(t) is plotted as filled circles as a function of time, t. X_{CHOL} = .83; $c_{TFB} = 5 \times 10^{-8}M$; membrane area: 6.1×10^{-3} cm^2. The solid trace is the regression line (slope = -1.66; intercept = -16.6; correlation = .999) computed for the data points shown.

---

[1] A clear and detailed discussion of the effects of unstirred layers on the time course of TFB current is given by Läuger and Neumcke (1973) and by Haydon and Hladky (1972).

Eq. (156) of Läuger and Neumcke (1973) or Eq. (A27) of Haydon and Hladky (1972)]. Therefore, most of the current seen at short times during the exponential decay must arise from translocation of TFB *adsorbed* at the membrane surfaces. This implies that assumption (18) of the theoretical section is obeyed for TFB to a good approximation so that the decay of the TFB current is simply a result of a concentration gradient built up at the membrane boundaries by translocation of TFB. The exponential decay of the current, which is in accord with the prediction of Eq. (22) as well as the polarization current that is observed when the membrane potential is returned to its resting value, both support the validity of this contention.

Initial permeabilities P as well as time constants of the initial decay of current τ are plotted in Figure 5 as a function of aqueous TFB concentration for membranes with increasing cholesterol content. The left-hand panel shows P's calculated according to Eq. (28) from the initial membrane current that was obtained from the zero-time intercept of the regression line through ln $I(t)$ versus t data (Figure 4). Dashed lines connect P's for membranes having the same composition. Increasing values of P's are for $X_{CHOL} = 0$, .5, .75, .91 (filled circles) and $X_{CHOL} = .83$ (open circles).

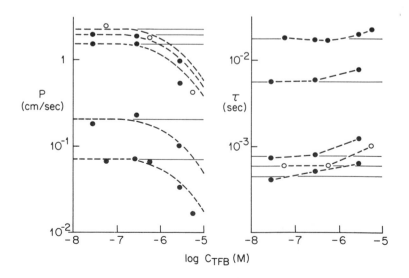

log C_{TFB} (M)

Within experimental error, P's are independent of the TFB concentration for $c_{TFB} < 10^{-6}M$. These values, shown in Figure 5 by thin solid lines, are also shown numerically in Table 1. At larger concentrations of TFB ($c_{TFB} > 10^{-6}M$) P's decrease by approximately the same factor for all values of X_{CHOL}. This range of complicated permeability behavior, not well understood, but also observed for phospholipids (Ketterer et al., 1971), probably results from decreased TFB

Table 1

X_{CHOL}	P (cm/sec)	τ (msec)	P_m/δ (sec^{-1})	$\gamma\delta$ (cm)	$\bar{P}_m$	$\bar{\gamma}$
0	.072	18	28	2.6×10^{-3}	1	1
.5	.21	5.6	89	2.3×10^{-3}	3.2	.9
.75	1.5	.78	640	2.4×10^{-3}	23	.9
.83	23	.60	830	1.7×10^{-3}	30	.7
.91	20	.46	1100	1.2×10^{-3}	39	.5

Fig. 5.   Left-hand panel, membrane permeability as a function of aqueous TFB concentration. Starting from the bottom, dashed lines connect data points for $X_{CHOL} = 0$, .5, .75, .91 (filled circles) and .83 (open circles). Solid lines show P's for $c_{TFB} < 10^{-6}M$. The dashed lines were drawn to the equation $P' = P(1/1+Kc_{TFB})$ with $K = 3.2 \times 10^5 M^{-1}$. Right-hand panel: time constant for the decay of the membrane current as a function of TFB concentration. Starting from the top, dashed lines connect data points for $X_{CHOL} = 0$, .5, .75, .83 and .91. Data points for $X_{CHOL} = .91$ are open circles; all others are filled circles. Solid lines show the value of τ's for $c_{TFB} < 10^{-6}M$.

activity, and is of only cursory interest here.[1]

The righthand panel shows τ's calculated from the slope of the regression line through $\ln I(t)$ versus t data. Dashed lines simply connect experimental data which are, in order of decreasing τ, for $X_{CHOL} = 0$, .5, .75, .83 and .91. Data points for $X_{CHOL} = .83$ are shown by open circles; all others are shown by filled circles. At low concentrations of TFB ($c_{TFB} < 10^{-6}M$) and within experimental error, the τ's are independent of c_{TFB}. The values shown in Figure 5 by thin solid lines, are also shown numerically in Table 1. For $c_{TFB} > 10^{-6}M$ the τ's increase slightly but significantly for all values of X_{CHOL}. As for P's, this range of complicated behavior is of only cursory interest.

Using Eqs. (28), (29), and (30) and the data of Table 1 it is now possible to calculate separately the translocation rates, P_m/δ, as well as the coefficients of surface partition, $\gamma\delta$, for membranes with an increasing cholesterol content. The results are summarized in the fourth and fifth columns of Table 1. The relative permeability of the membrane interior, $\bar{P}_m = P_m/P_m^{MO}$, and the relative partition coefficient, $\bar{\gamma} = \gamma/\gamma^{MO}$, are calculable directly from the fourth and fifth columns of Table 1, assuming that δ, the thickness of the adsorption layer, is independent of the lipid composition.

Figure 6 plots $\bar{P}_M$ (upper part, filled circles) and $\bar{\gamma}$ (lower part, filled circles) as a function of the lipid cholesterol mole fraction, X_{CHOL}. Comparison of the upper ($\bar{P}_M$) and lower ($\bar{\gamma}$) parts, both of which were drawn to the same

---

[1]*The decrease of P is possibly not real but apparently caused by a decrease in the concentration of free TFB resulting from the formation of TFB aggregates. The dashed lines in the lefthand panel which were drawn to the equation $P' = P(1/1+Kc_{TFB})$ with $K = 3.2 \times 10^5 M^{-1}$ and which are expected for the formation of impermeant TFB dimers in the aqueous phase, appear to account reasonably well for the decreased P's. A small surface potential, also expected for large surface concentrations of TFB, could explain the slight concomitant increase in the τ's for $c_{TFB} > 10^{-6}M$.*

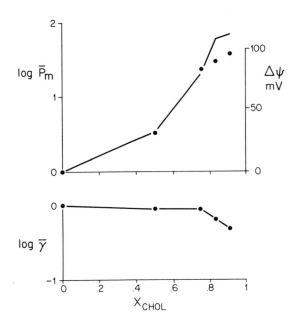

Fig. 6. *The influence of membrane composition on the TFB per-*
meability of the membrane interior (upper part) and
on the TFB surface partition (lower part). Filled
circles, plotted using the lefthand scales, show the
results of kinetic experiments. The solid line
drawn using the upper righthand scale, shows P_m's
that would be expected if the dipole potentials al-
tered only P_m and not γ. The scales are such that
58.7 mV corresponds to a ten-fold change in P_m.

logarithmic scale, shows that the predominant effect of lipid
composition is on the permeability of the membrane interior,
P_M, with only a secondary influence on the surface partition
γ of TFB. Thus, nearly all of the increase in the overall
TFB permeability P (Figure 2) results from an increased permeability of the *interior* of the membrane. That is to say,
X_{CHOL} alters mainly P_m and not γ.

The results of section 1 showed that increasing membrane cholesterol content increases the electrical potential
ψ_0 in the membrane interior. Since the partition coefficient

$\gamma = \exp[-(\mu_S^o + zF\psi_S)/RT]$ is practically unaltered for TFB, one must conclude that there are no changes in ψ_S, that is $\Delta\psi_S = 0$. The dipoles, from which the surface potential originates, must therefore be located toward the membrane interior relative to the adsorption plane of TFB. Consistent with this finding is the nearly exact agreement on the upper part of Figure 6 between $\log \bar{P}_M$ (left hand scale, solid circles) and $\Delta\psi_O$ (right hand scale, solid line) which is predicted from Eq. (12) assuming that all of the dipole potential is within the membrane interior, that is $\Delta\psi_O = \Delta\psi_M + \Delta\psi_S = \Delta\psi_M$ with $\Delta\psi_S = 0$ as before.

In summary, dipole potentials in monoolein-cholesterol membranes alter the permeability of the membrane interior but do not alter the surface partition for TFB. This implies that TFB is adsorbed *outside* the dipolar groups from which the potential originates.

DISCUSSION

The polar head groups both in monoolein, which has carbonyl and hydroxyl residues, and in cholesterol, which has a hydroxyl residue, are chemically inert and neutral. Bilayer membranes formed from these lipids are expected, therefore, to have a relatively nonreactive surface which has no net electric charge. The absence of surface charge originating from adsorption of ions or contaminants is further ruled out for these membranes by the observation that, aside from small activity effects, permeability is independent of the aqueous ionic strength both for lipophilic ions and for neutral carriers of ions (Szabo *et al.*, 1972; Szabo, unpublished observation). The increasingly positive electrostatic potential ψ_O detected by permeant ions in monoolein membranes of increasing cholesterol content must originate from oriented molecular dipoles. The existence of "dipolar" potential changes ($\Delta\psi_O$'s) assessed here from cationic and anionic permeabilities is not unexpected. Studies of monolayers spread at the air-water interface (Davies and Rideal, 1963) indicate that the compensation potential, which tends to measure a potential that exists at the hydrocarbon tails relative to the aqueous subphase, is about 90 mV more *positive* (~400 mV, Adam, 1968) for cholesterol than it is for monoolein (300-320 mV, Szabo and Eisenman, 1973; Hladky and Haydon, 1973). Since the bilayer can be viewed as two monolayers touching at the hydrocarbon tails, the $\Delta\psi_O$'s detected in bilayers could have been anticipated from the monolayer studies. It should be emphasized, however, that only direct measurements of $\Delta\psi_O$, as carried out here, can yield reliable estimates of the $\Delta\psi_0$'s that actually occur in bilayer membranes, since packing of the lipid molecules may be very different in bilayers and mono-layers.

The large changes in dipole potential point to the possible existence of large electrostatic potentials in the membrane interior. This absolute potential, ψ_0, could be determined in principle by comparing membrane permeabilities for ions which have the same intrinsic permeability λ, but opposite charge [see Eq. (15)]. As noted by LeBlanc (1970) TFB and TFP, which are practically identical species except for their charge, may be expected to have similar intrinsic permeabilities. Assuming that the λ's are equal for TFB and TFP, Eq. (15) and the permeability data of Table 1 yield 132 mV and 245 mV respectively as the value of the potential ψ_0 in the interior of membranes with $X_{CHOL} = 0$ and .91. Unfortunately, these estimates may not directly reflect ψ_0 since λ may be quite different for TFB and TFP [Eq. (14)] due to differences in the hydration energies of these ions. A reliable estimate of hydration energies would, in principle, allow an estimate for ψ_0, but at present the hydration energies of TFB and TFP are not known precisely enough for such an estimate.

Little is known about the precise location or identity of dipolar residues from which the dipole potential originates. Measurements of monolayer compensation potentials (Davies and Rideal, 1963) indicate, however, that the most likely origin of the dipole potential is in the lipid polar head groups and possibly in the water molecules that solvate them. For example, relocation of the hydroxyl residue from the 3β to the 3α position in cholesterol has the effect of decreasing the compensation potential from 350 mV to -100 mV (Adam, 1968). A convenient diagrammatic localization of the dipole potential is shown here in Figure 7. The lower part shows the head group (shaded) and hydrocarbon tail (light) of a cholesterol (upper) and of a monoolein (lower) molecule in the bilayer membrane. The top part shows a plausible potential profile that would be expected for a dipole potential originating from the hydrocarbon side of the polar head groups. Since there are no strongly dipolar groups in the membrane interior, the potential is expected to remain constant in the membrane interior.

The kinetic measurements of Section 2 have shown that TFB adsorbs at the membrane surface. This is not unexpected for a hydrophobic ion such as TFB which not only tends to be

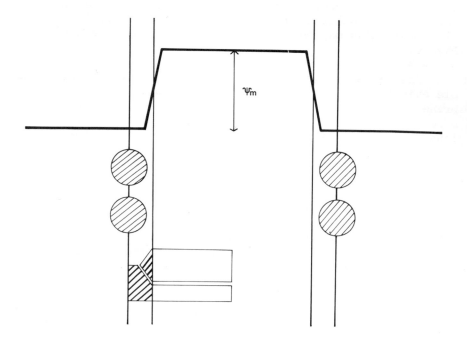

Ψ_m

*Fig. 7. Schematic representation of the bilayer membrane.
Double vertical lines show the region of polar head
groups; the area between these shows the hydrocarbon-
like membrane interior. The lower part shows the
polar head group (shaded) and hydrocarbon tail
(light) of a cholesterol (upper) and of a monoolein
molecule in the membrane. The top part shows the
potential profile expected for dipoles at the hydrocarbon side of the polar head groups. Shaded disks
show the most likely location of TFB adsorbed at the
membrane surface.*

expelled from the aqueous solution since it does not hydrogen
bond strongly with water molecules (Tanford, 1973), but also
tends to be expelled from the membrane interior since a large
electrostatic energy is required for an ion of its size to
enter the hydrocarbon-like membrane interior. The kinetic
measurements have also shown that increasing dipole potentials

enhance translocation across the membrane interior without any significant effect on the adsorption of TFB. Since the dipole potentials are believed to originate from the polar head groups, the adsorption plane for TFB must be located at the aqueous surface of the polar head groups. The middle section in Figure 7 locates TFB (shaded disks) in a way that is consistent with the result of the kinetic measurements.

Direct and carrier-mediated ion transport are similar since in both mechanisms a hydrophobic ion permeates the membrane. Because of this similarity, the influence of dipole potentials on the kinetics of carrier-mediated transport should allow one to locate the ion-carrier complex by a procedure similar to that carried out here for TFB. Preliminary experiments (Szabo, 1975) indicate that for such well known carriers as trinactin and valinomycin the ion-carrier complex is adsorbed at the aqueous side of the dipoles, in the same way as TFB.

Hydrophobic ions are excellent probes of the electrochemical properties of bilayer membranes. Although their use was restricted here to monoolein-cholesterol bilayers, hydrophobic ions are expected to yield similarly useful information on the electrochemical properties of bilayers formed from a variety of lipids (e.g., phospholipids). Furthermore, hydrophobic ions are expected to be extensively useful in investigating the double layer and dipole potentials created in bilayers by the adsorption of biologically active molecules.

ACKNOWLEDGMENT

This work was supported by grants from the USPHS (HL 16839), NSF (GB 30835), and the MDAA.

REFERENCES

ADAM, N. K. (1968). "The Physics and Chemistry of Surfaces," Dover, New York.

DAVIES, J. T., and RIDEAL, E. K. (1963). "Interfacial Phenomena," Academic Press, New York.

DeLEVIE, R., SEIDAH, N. G., and LARKIN, D. (1974). J. Electroanal. Chem. Interfacial Electrochem. 49, 153.

HALL, J. A., MEAD, C. A., and SZABO, G. (1973). J. Memb. Biol. 11, 75.

HAYDON, D. A., and HLADKY, S. B. (1972). Quart. Rev. Biophys. 5, 187.

HLADKY, S. B., and HAYDON, D. A. (1973). Biochim. Biophys. Acta 318, 464.

KETTERER, B., NEUMCKE, B., and LÄUGER, P. (1971). J. Memb. Biol. 5, 225.

LÄUGER, P., and NEUMCKE, B. (1973). In "Membranes" (G. Eisenman, ed.), Marcel Dekker, New York.

LÄUGER, P., and STARK, G. (1970). Biochim. Biophys. Acta 211, 458.

LeBLANC, O. H. (1969). Biochim. Biophys. Acta 193, 350.

LeBLANC, O. H. (1970). Biophys. J. 10, 94a.

McLAUGHLIN, S. G. A., SZABO, G., EISENMAN, G., and CIANI, S. M. (1970). Proc. Nat. Acad. Sci. U.S.A. 67, 1268.

SZABO, G. (1974). Nature 252, 47.

SZABO, G. (1975). Biophys. J. 15, 306a.

SZABO, G., and EISENMAN, G. (1973). Biophys. J. 13, 175a.

SZABO, G., EISENMAN, G., and CIANI, S. (1969). J. Memb. Biol. 1, 346.

SZABO, G., EISENMAN, G., McLAUGHLIN, S. G. A., and KRASNE, S. (1972). Ann. N. Y. Acad. Sci. 195, 273.

TANFORD, C. (1973). "The Hydrophobic Effect," John Wiley and Sons, New York.

THE EFFECT OF HIGH PRESSURE ON PHOSPHOLIPID BILAYER MEMBRANES

James R. Trudell

A study of the effect of inhalation anesthetics on the biophysics of synthetic phospholipid bilayers led us to the high pressure studies discussed in this paper.   We proposed this bilayer as a good model for the site of action of inhalation anesthetics.   We reasoned that if this bilayer is, in fact, a good model for the site of action of anesthesia, it must demonstrate all the phenomena known about anesthesia.

One such phenomenon which we decided to use as a test of our model was the pressure reversal of anesthesia (Lever, 1971).   An example of pressure reversal is the introduction of an anesthetic gas into a chamber containing several mice, observing the resulting loss of righting reflex in the mice, applying 100 atmospheres of helium pressure to the gas in the chamber, and observing a restoration of activity and righting reflex.

We reasoned that if our model was a good one, the increase in fluidity in the bilayer that we observed, due to the action of anesthetic drugs, likewise must be reversed by high pressure.   In a series of experiments we demonstrated that this was so (Trudell, 1973a).   We further demonstrated that the effect of high pressure is not to exclude the small

anesthetic molecule from the bilayer, but rather to reorder
the bilayer around the anesthetic molecule (Trudell, 1973b).

We then turned to study how a modification of the fluidity of the neural membrane might modify the action of proteins
which we think are essential to modulating nerve conduction.
One such possibility was a modification of the lateral phase
separation properties of a nerve bilayer. Studies by C. Fred
Fox (1972) on bacterial mutants, and Shimshick and McConnell
(1973) on phospholipid bilayer systems, have shown the importance of phase transitions in the membrane lipids to the
functioning of membrane-solvated proteins. Such studies
suggested to us the possibility that anesthetic-induced phase
transitions might be important to the mechanism of anesthesia.
We thus studied the effects of anesthetics on phase transitions and indeed observed a lowering of phase transition temperatures by inhalation anesthetics.

We then studied the effect of high pressure on phase
transitions in order to observe the possible antagonism
between high pressure and inhalation anesthetics on these
systems. A pressure of 136 atm was shown to cause a 3.0° ele-
vation in the phase transition temperature of dipalmitoylphosphatidylcholine bilayers. This change in phase transition
temperature is very close to that predicted from the Clapeyron
equation by substituting the value we observed for ΔT as well
as the ΔV for a phase transition which has been calculated
from other techniques by other workers. Thus, it would
appear that the pressure effect on phase transitions is pre-
dictable using theories of first-order phase transitions
(Trudell, 1975).

Quite apart from the use of high pressure as a tool to
study the mechanism of anesthetic action, high pressure is
important to understanding the behavior of humans diving to
great depths, deep diving fish, as well as other organisms
subject to extreme environments of high pressure. For example, I have described the electron paramagnetic resonance
(EPR) experiment which demonstrated that the internal fluidity of a membrane decreases upon application of 100 atm of
pressure. This change is sensed by proteins solvated in the
bilayer as a higher internal viscosity as well as a change
in the lateral compressibility of their surroundings. Such
changes might be expected to influence the function of these
membrane-solvated proteins. In addition, our work with the
effect of high pressure on raising the phase transition temperature of either pure or mixed phospholipid bilayer systems

suggests that high pressure might have a more profound effect in certain biological systems. According to the suggestions of Fox and McConnell, some biological systems tend to maintain their membrane lipid regions in a condition in which there is some fluid and some solid phase present at all times. Such a phase separation would allow for maximum lateral compressibility of the surrounding bilayer matrix. They suggest that such lateral compressibility is important to the function of membrane-solvated proteins.

If application of high pressure raises the phase transition temperature of all of the phospholipids in such a membrane, then at the same temperature more of the lipids in the membrane will be below their phase transition temperature. That is, more or all will be in the gel phase, and the essential fluid-gel equilibrium will no longer exist.

Figure 1 illustrates the insertion or expansion in volume of a membrane-solvated protein. In this drawing the expansion or insertion of a protein is schematically illustrated as a wedge being forced into the surface of the phospholipid bilayer. The bilayer on the left side is shown to have gel-phase phospholipids near the point of wedge insertion. In the right diagram, it is seen that the phospholipid matrix accommodated the insertion of the wedge by transforming some

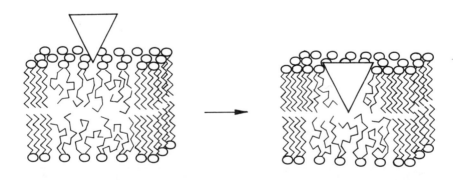

Fig. 1.   The insertion or expansion of a protein (wedge) in a phospholipid bilayer.

of the fluid phase phospholipids into the more condensed solid
or gel phase, thus making room for the volume of the wedge.
This is a way in which the lipid matrix which has a fluid-gel
equilibrium is able to allow for protein expansion or conformational change. If such a lipid were all in the gel phase
already, such accommodation would be impossible. This model
would predict that at some point following the application of
high pressure, a membrane would exist in the all-gel phase
and lose its lateral compressibility.

Figure 2 is a diagram of the phase transition of a pure
dipalmitoylphosphatidylcholine bilayer. The internal fluidity of the bilayer is plotted against temperature. The curve
depicting the phase transition is shown at one atmosphere,
69 atm, and 136 atm.   It is seen that both the pretransition
point in the region of 33°, which is thought to involve the
rearrangement of the phospholipid head groups, and the major
transition at 43°C, which represents the fatty acid chains of
the phospholipids going from a gel to a fluid state, are
influenced by the application of pressure. The effect of

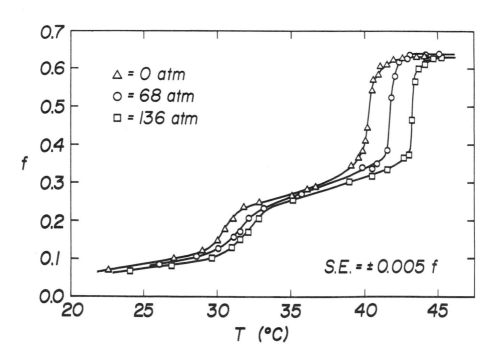

Fig. 2. Phase transition of a pure dipalmitoyl-phosphatidyl-
choline bilayer. f = internal fluidity.

pressure on membrane fluidity is best seen if one observes a point on an isothermal line drawn vertically at 43°C, which is the center of the phase transition for the 68 atm curve. It will be noted that at atmospheric pressure on this isothermal line the bilayer is nearly all in the fluid phase (high f). As one follows the isothermal line to connect with the 136 atm curve, however, it is seen that now the bilayer is nearly all in the gel phase. We have also shown this effect to be true for mixed systems of dipalmitoyl- and dimyristoyl-phosphatidylcholine.

We are attempting to correlate the possible change in lateral phase separation behavior in a membrane with the site of action of inhalation anesthetics. We have shown that this anesthetic action is antagonized by high pressure. As a subject for our further study, we intend to explore the possibility that part of the effect of high pressure on organisms is caused by the shift of lateral phase separation temperatures.

REFERENCES

FOX, C. F. (1972). In "Membrane Molecular Biology" (C. F. Fox and A. D. Keith, eds.), pp. 145-161. Academic Press, New York.

LEVER, M. J., MILLER, K. W., PATON, W. D. M., and SMITH, E. B. (1971). Nature 231, 368.

SHIMSHICK, E. J., and McCONNELL, H. M. (1973). Biochemistry 12, 2351.

TRUDELL, J. R., HUBBELL, W. L., and COHEN, E. N. (1973a). Biochim. Biophys. Acta 291, 335.

TRUDELL, J. R., HUBBELL, W. L., COHEN, E. N., and KENDIG, J. J. (1973b). Anesthesiology 38, 207.

TRUDELL, J. R., PAYAN, D. G., CHIN, J. H., and COHEN, E. N. (1975). Proc. Nat. Acad. Sci. U.S.A. 72, 210.

Symposium Lectures   1970

Protein Structure   Peter von Hippel

Ion Permeability   Lester Packer

High Temperature   L. Leon Campbell

Subzero Temperatures   Peter Mazur

Hydrostatic Pressure   Richard Y. Morita

Effects of Radiation   John Jagger

Low Water Concentration   Harlyn O. Halvorson

High Salt Concentration   Donn J. Kushner

The Meeting also included the presentation of 40 short papers.
The Proceedings were not published.

Symposium Lectures   1972

Proteins   Irving M. Klotz

                                            Rufus Lumry

                                            Rodney Biltonen

Membranes   M. R. J. Salton

                                            Youssef Hatefi

                                            Joseph Steim

Autotrophy   Howard Gest

                                            Harry D. Peck, Jr.

                                            A. N. Glazer

The Meeting also included the presentation of 39 short papers.
The Proceedings were not published.

Subject Index